当敬爱的朋友您掀开此书时，欢迎进入我们深藏内敛的设计世界，
并在此获得阅读的趣味和分享我们的喜悦。
你们的欣赏、共鸣、鼓励，是我们得以勇往直前的源动力。
献给我们挚诚的朋友和亲人们。

Sincerely to our beloved friends and families...
When you open this book, we expect you to step
into the design world from our restrained heart,
to enjoy the fun of reading, and to share our joy.
Your appreciation, empathy and encouragement
would always motivate us to move forward.

若不是耶和华建造房屋，建造的人就枉然劳力。若不是耶和华看守
城池，看守的人就枉然警醒。

—《圣经》诗篇127章1节

Except the LORD build the house, they labour
in vain that build it: except the LORD keep the
city, the watchman waketh but in vain.

Psalm 127: 1 in Holy Bile

SUNNY & ROBIN

研行十载
——SNP 实践与回顾 第二辑
FROM PRACTICE TO COLLECTION
-SNP DESIGN RECORD VOL.2

广州市唐艺网络科技有限公司 策划
广州尚逸装饰设计有限公司 编著

中国林业出版社
China Forestry Publishing House

CONTENTS 目录

这是个需要心平气和、坚持理想、重拾设计价值观的年代。设计，不仅是解决问题，更是美化生活，创造未来。多一份诚恳，就多一份品质专注；多一份执着，就多一份细节；多一分热爱，就多一分超越；多一分品德，就多获得一分尊重。我用我自己的身体与思考，坚持实践着这一真理，并希望给我身边的人带来这如同马太福音般的影响。由此，我们不得不去重新思考下有关设计的两个关键词，"专业"与"大师"。

This is an era which demands the calmness of heart, the perseverance to dreams, and the embrace of design value. Design is not only about problem-solving, but more about beautification of life and creation of future. Sincerity means providing better quality; persistence means paying attention to details; passion means going beyond boundary; virtue means earning more respect. I use my body and soul to persistently practise the truth, and hope I could influence other people like the Gospel of Matthew. Thus, we have to ponder on two key words in design: professionalism and masters.

004　序言 PREFACE
步履不停，友谊长青
Non-Stop Progress, Everlasting Friendship
虚与实
Spiritual And Actual

008　天伦办公室项目
SNP new office-Talent Holdings building

024　新东方风格 NEW ORIENTAL STYLE
广州保利天悦6号楼顶层复式、南京富力湿地会所、淮安金奥国际销售中心、沈阳保利茉莉公馆销售中心、天津保利天汇销售中心
6# Poly Grand Penthouse Guangzhou, Club Of R&F Wetland Park, Nanjing, Sales Center Of Kingtown International Center, Huaian, Sales Center Of Poly Jasmine, Shenyang, Sales Center Of Poly Grand Influx, Tianjin

072　地产系列 THE REAL ESTATE SERIES
太原保利西江月、福州保利天悦、广州龙璟山、北京保利和光尘樾、厦门保利叁仟栋
Poly River Coast, Taiyuan, Poly Grand Mansion ,Fuzhou King's Mountain, Guangzhou, Poly Palace Of Light, Beijing, Poly Costal Mansion, Tongan

154　地标建筑 LANDMARKS
三亚保利财富中心、广州保利国际广场、广州保利天幕广场、广州华南国际港航服务中心、珠海中冶盛世国际广场
Poly Wealth Center,Sanya, Poly International Plaza, Guangzhou, Poly Skyline Plaza, Guangzhou South China International Port And Waterway Service Center, Guangzhou, Mcc Honour Plaza, Zhuhai

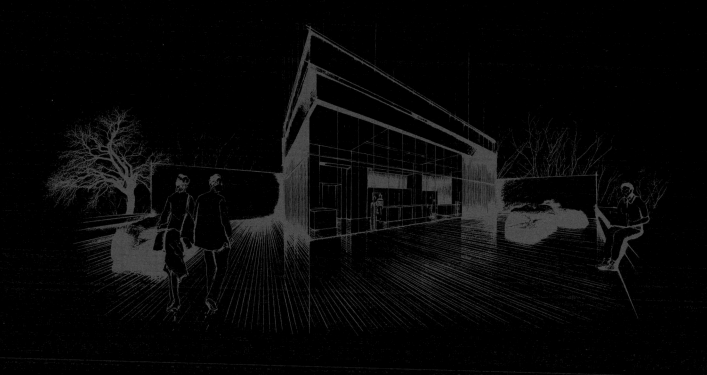

246 商业空间 COMMERCIAL SPACE

南京& 上海诺阁雅酒店、南京保利中央公园销售中心、广州保利天汇会所、深圳海上世界沙龙会所、启迪协信杭州科技城展示中心、上海协信集团企业会所、深圳贝骊洛生活美学馆
Neqta Hotel, Nanjing & Shanghai, Sales Center Of Poly Center Park, Nanjing, Club Of Poly Grand Influx, Guangzhou, Art Salon In Sea World, Shenzhen, Exhibiton Center Of Tusincere Science City, Hangzhou, Sincere Corporate Club, Shanghai, Best & Real Living Zone, Shenzhen

310 别墅 VILLA

重庆保利江上明珠依山别墅、北京保利首开天誉别墅、香江海滨私人别墅
Villa Of Poly Real Estate, Chongqing, Villa Of La Vita E Bella, Bejing, Private Villa On Pearl River

346 现代主义 MODERNISM

重庆保利江上明珠公寓、合肥保利海上明悦销售中心、莆田保利拉菲公馆销售中心、北京和锦薇棠销售中心、清远保利春晓接待中心
Loft Of Poly Real Estate, Chongqing, Sales Center of Poly Sea Bright, Hefei, Sales Center Of Poly Lafite Mansion, Putian, Sales Center of Poly Belle Ville, Beijing, Reception Center of Poly Spring, Qingyuan

383 后记 POSTSCRIPT

源自生活，追求卓越
From Life To Life
从零到壹的视界
From Zero To One

序言 / PREFACE

步履不停，友谊长青

盛宇宏
中国建筑学会理事
羊城设计联盟会长
汉森伯盛国际设计集团董事长/总建筑师

吾与王赟（Sunny）一直是同奋斗在华南区设计圈的同行，他年龄比我小一轮，都在年轻时代投身创业，充满了激情与梦想，也经历了艰难困惑。Sunny年轻、有冲劲、十分努力，也常常有很多有冲击力的点子。时光匆匆，如今他和伙伴们创建的尚诺柏纳（SNP）也已经走过了10年之路，可喜可贺！这次Sunny邀我写序，意外之余，倍感荣幸，也有很大压力。

SNP的发展非常迅猛，近几年常见Sunny和他们公司的作品。过去10年，国内市场的消费观念和方式不断变化，人们对室内设计的要求越来越高。建筑设计和室内设计都纷纷打破以往方盒子的限制，使功能空间更加多变、丰富、并且明晰。多元、复合、智能、环保、时效等要求必然成为未来发展的趋势，而SNP的作品，正是在不断的做着这些方面的解读和验证。

设计师常被人誉为灵感与创意的职业，创意、灵感可以来源于想象力，但身在其中，我们知道，更多的是来源于对生活的积累与沉淀，对文化的洞察与体验。这几年一直在观察和学习SNP创作的空间作品，看到了他们的不懈努力和长足的进步。建筑师和室内设计师一样，设计都需要"从人出发、以人为本"，SNP的作品从结构到细节都充满了人文关怀，第一时间先符合生活的需要，处处体现生活的痕迹；在关怀人的基础之上，他们不忘提升与引领审美标准，带入先进的设计理念，比如近几年SNP的作品：北京保利和光尘樾、三亚保利财富中心、广州保利天幕广场、协信集团会所、中冶盛世国际广场等等，都相当让人惊艳！在细节之处，有许多让人回味和思考的地方。可见其团队在设计过程中思维之严谨，并以客观科学的思维作为基础，以为人们提供更美好的生活为目标，运用团队的专业技术再去考虑外显的美学问题。

2016年起，我担任羊城设计联盟（羊盟）的会长，Sunny成为副会长，我们一起共事，因为羊盟日渐熟悉，这其中更感觉到其谦逊有礼和求知上进；他后进自谦，但他带领下的设计团队每一年都在飞速进步，SNP已成大家，令人相当佩服！与Sunny一起作为羊盟理事会的核心领导团队，我们合作亲密无间，广州乃至华南区的民间设计力量需要走出去，需要发扬光大，这是我担任羊盟会长所设立下的奋斗目标之一。Sunny建言力行，拿出许多宝贵时间，为推动羊盟的向前发展贡献其力量。在此我为有这么一位好弟兄深感荣幸和快乐，也寄望羊盟在以后的发展中，能与Sunny等众多好伙伴携手共进。

SNP作为一家民营的室内设计公司，步履不停，走过10年光景，如今愈发壮大，已可预见SNP的未来更加美好！在此祝愿王赟和SNP，设计水平不断创新提升，不但在华南区引领先锋，在全国大有作为，未来也能在世界的舞台上，让大家见识中国的设计力量！共勉之。

序言 / PREFACE

NON-STOP PROGRESS, EVERLASTING FRIENDSHIP

Eric Shing
Committee Of Architectural Society of China
Chairman Of Yangcheng Design Alliance
Chairman Of Shing & Partners International Design Group

Sunny and I are peers in the design industry of southern China. He is a dozen years younger than me full of passion and dreams, plunging into the entrepreneurship at a young age and weathering through difficulties and hardships. He is young, diligent, and full of power with many creative ideas. Time flies, SNP founded by his partners and him has witnessed the first decade. Congratulations are in order! That Sunny has invited me to write the preface is a surprise and honor for me, and of course quite a stress.

SNP is under a rapid development with productive works over the recent years. The last decade has witnessed continuous changes on the ideas and ways of consumption in the domestic market as well as higher demands for interior design. Both architectural and interior designs have jumped out of the traditional "cube house" and made the space more flexible, versatile, and distinct. The future trend is to be diverse, compound, intelligent, environment-friendly, efficient, and etc., which is exactly what works of SNP aim for.

Designers are often associated with inspiration and creativity which may originate from imagination. But in fact, it is more likely to originate from life experience and cultural understanding. The works of SNP in recent years demonstrate their unremitting efforts and significant progress. Both architectural and interior designers should put people first, which can be reflected from SNP's works from structure to details—meeting the needs of life. Besides, they keep improving and guiding the aesthetics standard as well as introducing advanced design ideas, which can be seen in SNP's amazing works over the recent years: Beijing Poly Palace of Light, Sanya Forum Complex, Guangzhou Poly Skyline Plaza, Sincere Group Club, MCC Real Estate Grand Plaza, and etc. Many details are worth reflecting and thinking. As one can see, this team applies rigorous, objective and scientific thinking as well as professional skills in design with an aim to provide better life for people, and then aesthetics follows.

Since 2016, I have served as the chairman of Yangcheng Design Alliance and Sunny the vice chairman. We work together and get better acquainted with each other day by day, and I get to know him to be even more modest, polite, and eager to make progress. He seldom talks about his achievement but the design team he leads achieves rapid progress continuously and SNP has become an admirable top firm in the industry. We make a good team in the council's core leadership of Yangcheng Design Alliance. The folk design of Guangzhou and the rest of southern China needs to go global, which is one of the aims set by me as the chairman of the Alliance. Sunny puts forward lots of constructive advice and acts on it, contributing a lot of time and efforts to advance the development of the Alliance. Hereby, I am honored and pleased to have such a good friend and brother and hoping to stride forward with him and other good partners in the Alliance.

SNP, as a private interior design firm, has gone through its first decade with non-stop progress, getting stronger with a promising future. Hereby, my sincere wishes go to Sunny and SNP for incessant innovation and improvement in design, and being a vanguard not only in southern China, but nationwide and even worldwide—let the world get impressed by Chinese design!

序言 / PREFACE

虚与实

关鸣
广州绿色之春文化公司创始人
"思想沙龙"创始人、建筑文化布道者

2018年，是SNP成立第10个年头，首先祝贺SNP十周年顺利开展！

十周年，对于一个成长中的设计企业来说太重要了，这是第一个里程碑，值得记载、回望、总结、反思；值得与公司内部同事们分享公司成长的经历，探讨未来发展之路如何走；值得与外界同行分享与交流设计企业的价值与责任。

这些看似比较"虚"的东西特别需要精心营建，在企业遇到低谷或者处于困难时，这些"虚"的东西可以支撑一个设计企业在逆境中坚定前行。

我在与SNP的合伙人王赟和王小锋多年的交往中，感受到他们身上的一些可贵之处：好奇心、浪漫、孩子气。也许这些特点在很多人眼中看是不成熟、不稳重的特点，怎么能成大器？怎么能做企业领导者？但是在我的眼中，却是创意和设计企业掌门人应该拥有的非常重要的性格特点，甚至是卓越的设计企业掌门人亟须的从业要素。

SNP是带了些学院派特点的设计企业。学院派未被很多设计同业所推崇，通常认为它的做法不合时宜，如何能服务市场，服务客户？

SNP在管理架构中，有研发部，有研究员。他们的工作是为公司其他生产部门提供设计理论支持、数据成果、综合信息的支持。这个部门经过数年的积累，已经成为SNP"虚"的部分中最有价值的基石。

还有一个例子给我印象深刻，那时SNP刚搬入白云大厦办公室。不久，公司内部举办的设计方案评比活动，记得有十多个不同的项目组参加，参加评比的项目都制作了精美的展板、有些还有模型，公司的设计师可以自由投票，发表自己对项目的不同看法。当时，SNP的业务已经很繁忙，我问王赟为什么还要举办这样的公司内部设计比赛？他说公司项目多，很多项目小组之间都不太清楚互相在做什么项目，也不清楚为什么有的项目做的好，获得市场和很多业主的好评。公司内部的设计比赛，像是一个内部的专业交流活动，部门相互之间可以学习、交流，起到相互促进的作用。

最近，我和Gracie再次被邀请去SNP新搬迁不久的办公室（天河区天伦控股大厦）。这次，他们的室内绿坪会议+阅读+会客的"多功能区"已经完成。虽然布局紧凑，但在寸土寸金的CBD办公室里有这样一片"绿洲"，让人感觉眼前一亮。我们在这片"绿洲"上参与了王赟工作室与广州美院杨一丁以及林红老师带领的研发设计小组的方案讨论会，这是一个典型的"研发"设计项目讨论会。

SNP联合高等学府建筑学院，与老师和学生共同研发、设计，以现实的商业项目作为研发课题，引起双方实践方式的碰撞，而非简单的业务外包。这样大胆的尝试，需要投入耐心与时间，最终的效果如何现在还不得而知，不过，我很赞赏这样的做法。

SNP走过第一个10年，代表一个设计企业已经具备一定的抗压抗打能力。祝愿SNP在下一个10年的发展中，迈向一个更高的台阶，成为本土设计行业里备受尊重的综合性设计领袖。

序言 / PREFACE

SPIRITUAL AND ACTUAL

Tom Kwan
Founder of Guangzhou Green Spring Media Co., Ltd.
Dean,Cheung Kong School of Art and Design,Shantou University

The year of 2018 marks the 10th year since the establishment of SNP. First of all, congratulations on the 10th anniversary!.

The 10th year is very important for a growing design firm. This is the first milestone. It is worth documenting, reviewing, summarizing and reflecting; it is worth sharing the firm's growth experience with colleagues and exploring the future development; it is worth sharing and communicating the value and responsibility of design companies with our peers.

These seemingly "spiritual" things need to be run carefully. When a firm is in trough or in difficulty, these "spiritual" things can support a design firm to move forward in the face of adversity.

During my years of association with SNP partners Wang Yun and Wang Xiaofeng, I found out something about them: curious, romantic, and childish. Perhaps these characteristics indicate immaturity and instability in the eyes of many people. They might think: how could they succeed? How could they be business leaders? But in my eyes, these are very important personalities that the heads of those creative and design firms should have, and even the essential features that the head of an excellent design firm should have.

The SNP is a design firm with some academic features. However, academism is not highly recommended by many peers of the same trade. It is generally considered that academism is out of date and unable to serve the market or clients.

SNP has a R&D department in the management structure, whose job is to provide design theory support, data results, and comprehensive information support for other production departments. After years of accumulation, this department has become the most valuable cornerstone of the "spiritual" part of SNP.

There is another example: Not soon after SNP moved into the Baiyun Building office, it held a design scheme competition within the firm.

I still remember that a dozen of project teams participated in the competition. They had made beautiful panels and some even made models. SNP's designers could vote freely and express their own opinions on different schemes. I asked Wang Yun with so many projects at hand why SNP still held such a design competition. He said that there were many project teams in the firm, and they were not very clear about what other teams were doing and why some projects were well-received by the market and many clients. The competition was more like a professional exchange so that different project teams could learn from and communicate with each other and eventually achieve mutual progress.

Recently, Gracie and I were once again invited to the newly relocated office of SNP (Tianlun Holdings Building in Tianhe District). This time, the "mufti-functional area" of meeting + reading + reception on the indoor lawn has been completed. The layout is compact, but such an "oasis" in the CBD office is really refreshing. In this "oasis", we participated in the seminar of the R&D design team led by Wang Yun's studio and Yang Yiding and Lin Hong from Guangzhou Academy of Fine Arts, which was a typical "R&D" design project seminar.

SNP worked with teachers and students from the Academy and used real business projects as R&D topics, causing collisions of practical approaches between the two sides, rather than simple business outsourcing. Such bold attempts require patience and time. The final result is still unknown, but I appreciate this approach.

SNP went through the first 10 years, representing that the firm already has a certain ability to cope with pressures. I hope SNP will move to a higher level in the next decade and become a respected and comprehensive leader in the local design industry.

天伦控股大厦19层办公室
SNP NEW OFFICE -TALENT HOLDINGS BUILDING

关键词：重整、不规则、想象空间
Keywords: renovated, irregular, imaginary

建筑设计 / 华南理工大学建筑设计研究院设计五所
室内设计 / 尚诺柏纳空间策划联合事务所
楼层面积 / 1,596 平方米

2017年，SNP应团队和业务的规模发展情况，决定搬到坐落于广州天河林和片区的天伦控股大厦。大厦建筑以平行四边形塔楼以及不规则多边形裙楼形态来对话周边环境，富有时尚和活力的姿态将给人们带来更广阔的想象空间。

生活艺术的开端

前台背景墙上的一幅长卷，在灯光的照射下形如一缕烟火或一抹飘云。它由细密的钉子排列而成。金属的冷硬以别样的手法则形成柔美舒畅的图形，是艺术带给我们的二元辩证和灵感。放置在窗边的一角布艺沙发供客人休憩。生活的艺术，就在于细节和无声处。我们希望客人在等候的间隙也能感受到体贴的关怀。

"艺术源自生活，又高于生活。"源自生活的正是SNP一直以来耕耘的理念所在，而追求卓越则是我们的目标。我们是从事空间设计的设计师，也是实践生活艺术的设计师。

In view of the development of business and team size, SNP decided to relocate to the SNP New Office -Talent Holdings Building at the Linhe Block, Tianhe District, Guangzhou in 2017. With a parallelogram tower building and an irregular polygon skirt building, it appears fashionable and vigorous with infinite possibilities.

Start of artistic life

The scroll against the lighting on the background wall of the reception is like a wisp of smoke or a touch of cloud, made up of closely-distributed nails. In this way, the cold metal turns into a mild pattern—a credit to art. By the window lies a fabric sofa for guests to rest. The art of life lies in details and subtleness. We hope our guests can be carefully tended even during waiting.

"Art is from life and beyond it." Originating from life is our philosophy and striving for excellence is our goal. We are designers working on interior design as well as the art of life.

艺术长廊

连接各个办公区的即是一条横跨楼层的走廊。最新的公司资讯、历年的经典项目及所获的奖项都在这里重现和展示。中岛吧台吸引同事们在此互动交流和小憩。充满人气和活力的长廊，是连接彼此的重要桥梁。

Corridor of art

What connects the offices is a corridor linking different sections, where up-to-date information of the firm, classic projects over the years, and awards are displayed. The island-shaped bar attracts colleagues to come over for rest and talk. The popular and vigorous corridor is an important bridge linking one and another.

❶ 前台背景墙寓意着活力无限的城市脉搏。
The reception backdrop symbolizes the vibrant city pulse.
❷❸ 长廊连贯办公室的东西侧，是公司重要的形象墙。
The gallery as an important image wall of the company is adopted to connect the east and west sides of the office.
❹ 多个精巧的项目模型。
Multiple ingenious project models.

交流序列

"灵感"大会议室预留了多种布置方案以应对不同的会议场景。生趣盎然的墙纸临摹了葱绿的丛林，给人生机勃勃的意象。窗前的阅览区由几套高脚桌椅点缀而成，闲聊或阅读期间也能享受窗外的景色。造型独特的灯饰为这片空间的照明系统带来惊喜。每个会议区域对应不同规模、不同主题的会议场景，独特的命名彰显巧思。生活的种种场景，也是由"沟通"而生。无论是分享知识的讲座沙龙，还是制造欢乐的精彩活动，充满交流互动的办公环境，才能有效促进同事们的工作效率和热情。

Interactive sequence

The conference room "Inspiration" reserves multiple layout schemes for different meetings. Its wall paper with verdant forests delivers an air of vitality. The reading area consists of several sets of bar stools before the window, allowing people to enjoy views while chatting or reading. The lighting with an unusual design brings surprises. Different meeting areas with ingenious names correspond with meetings with different scales and themes. Life settings are also generated by "connection". Be it salons for information sharing or activities for fun, an interactive office environment is of great help to promote efficiency and enthusiasm.

 MEETING ROOM 灵 感

 VIDEO CONFERENCE ROOM 光 圈

 MEETING ROOM 知 影

 LIBRARY 视 界

 MEETING ROOM 神 韵

 MEETING ROOM 云 端

① 灵感大会议室是举办重要活动的场地。
The inspiration meeting room is a venue for holding important events.
② 重要的奖项均集中在中央走廊的东侧。
Significant awards are gathered on the east side of the central corridor.
③ 视频会议室采用清新自然的墙纸营造轻松愉悦的交流氛围。
Fresh and natural wallpapers are adopted in the video meeting room for creating a relaxing and cheerful atmosphere.
④ 各会议室的VI标示。由上到下分别命名为分别是灵感大会议室、光圈视频会议室、知影区域办公室、视界区域办公室、神韵视频会议室、云端区域办公室。
VI marks for all meeting rooms. They're named inspiration meeting room, aperture video meeting room, charm video meeting room, video office, vision office, and cloud-end office, from top to bottom.

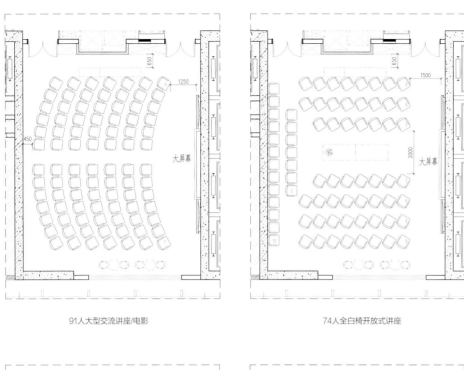

91人大型交流讲座/电影　　　74人全白椅开放式讲座

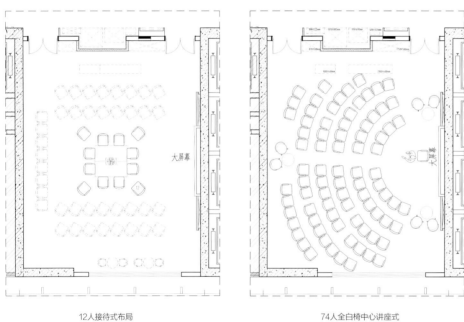

12人接待式布局　　　74人全白椅中心讲座式

多功能、多场景、多融合

作为一家为建设美好城市而付诸努力的进取学习型公司，我们欢迎各界朋友来和我们作经验分享交流。基于这一理念，我们设置了多功能综合区，它将作为一个开放式沙龙空间，具备活动接待、咖啡沙龙、藏书阅览等功能。每个人都可以在这个舒适的环境中获得一丝放松，于交流之中交换创意和灵感。

Multi-functional, multi-setting, and multi-blending

As a progressive company endeavoring for a better city, we welcome friends from all walks of life to share and communicate with us. To this end, we have set up a multi-functional area as an open salon with functions of reception, café, library, and etc. Everybody can ease up in this cozy environment, exchanging creative and inspirational ideas.

新东方风格
NEW ORIENTAL SERIES

东方情怀，西方格调。新东方代表现代艺术与传统文化融合的尝试。
我们用多个颇具东方韵味的作品，回应这股东西方交流碰撞的美学思潮。

Oriental feelings are combined with Western style. New oriental series are attempts to integrate modern art with traditional culture. We use a number of works with oriental charm to interpret the aesthetics from the eastern and western exchanges.

广州保利天悦 6 号楼顶层复式
南京富力湿地会所
淮安金奥国际销售中心
沈阳保利茉莉公馆销售中心
天津保利天汇销售中心
6# Poly Grand Penthouse, Guangzhou
Club Of R&F Wetland Park, Nanjing
Sales Center Of Kingtown International Center, Huaian
Sales Center Of Poly Jasmine, Shenyang
Sales Center Of Poly Grand Influx, Tianjin

一江之水，见证历史的变迁。广州琶洲湾新城改造建设进程最新实景。
The river witnesses historical changing. The latest pictures of the renovation process of Guangzhou Pazhou Bay New Town.

广州保利天悦6号楼顶层复式
6# POLY GRAND PENTHOUSE, GUANG

关键词：中西融合、奢华、一线江景
Keywords: fusion of East and West, luxury, river view

建筑设计 / 绵博建筑师事务所私人有限公司
景观设计 / 贝尔高林国际（新加坡）私人有限公司
建筑面积 / 1,850,000 平方米
委托范围 / 硬装及软装设计
委托面积 / 511 平方米

当东方遇上西方，当中国遇上意大利，当文化加上构思，便汇集成一处奢华之所。岭南文化元素装饰与意式家具的相遇，正是东西文化的交互，亦是设计师对奢华空间的创新尝试。本案毗邻珠江，坐落于豪阔江景客厅，南北通透双阳台，楼顶独家天际泳池，一城江色尽收眼底。

外有城市美景，内部如何打造符合城市精英的奢华住所。三层复式住宅，设计师以祥云图纹作为主要的装饰点缀，玉石的冷艳，皮质家具的装点，打破豪宅沉闷的印象。中式与西式的配饰交替在空间当中碰撞融合成中西韵味兼具的生活之所。

The combination of Lingnan cultural decorations and Italian furniture is an interaction between eastern and western cultures as well as the designer's bold attempt at luxurious houses. Next to the Pearl River, this building hosts spacious living rooms with full river views, dual terraces from south to north, and a rooftop swimming pool, enabling one to appreciate the river view to the fullest.

A luxurious residence for urban elites. This three-story property uses auspicious clouds as the main decoration, along with the cool and beautiful jades and leather furniture, successfully breaking the stereotype of luxurious residences being dull and monotonous. The fusion of oriental and occidental decorations makes it a living place accommodating both eastern and western charm.

❶❷ 设计师将东西文化元素融入客厅。
The designer combines the cultural elements of the eastern and western cultures into the living room.
❸❹❺ 首层各空间立面图。
Space elevation on the first floor.
❻ 顶层复式首层平面图。
The penthouse first floor plan.

客厅中摆放着宽长的沙发与散落式的独立沙发，可供多人相聚亦满足居住者平日的家庭聚会。软装配色以金色和紫色为主，金色的奢华与紫色的高雅相互衬托着空间的高贵。天花以祥云做浮雕，寓意对生活的美好祝愿，地毯的花纹与天花上的浮雕相呼应，中式的书法镶嵌于金色的画框中，不经意间展示出中式风韵。

There are long couches and scattered single sofas in the living room for parties or daily family gatherings. Their colors feature in luxurious gold and elegant purple, which complement one another and deliver an air of nobility. The ceiling uses auspicious clouds for embossments as a wish of good life. The patterns on the carpet echo the embossments of the ceiling. And Chinese calligraphy is displayed within golden frames. All of these naturally reveal the charm of Chinese culture.

- ❶❸❹ 设计师巧妙的将岭南文化的精髓融入到生活空间当中。
  The designer cleverly integrates the essence of south of the Five Ridges culture into the living space.
- ❷ 餐厅手绘图。
  The restaurant sketch.
- ❺❼ 雕琢着祥云图案的阶梯。
  The ladder pattern carved with auspicious clouds.
- ❻ 设计师用手绘图呈现对中空位置的构想。
  The designer presents the concept of hollow position by hand drawing.

走至餐厅，巨大的水晶吊坠点亮空间，圆形的托盘与餐桌暗喻世界皆为一体。楼梯连接着上下两层的空间，以玉石为圆形的基底构建，侧面雕琢着祥云图案。建筑中，大面积的玻璃幕墙激发了设计师的想法，用水晶吊饰组织成祥云与圆月的图案，填满中空位置的空缺。

Walking into the kitchen, a huge crystal chandelier comes into sight, whose round base along with the round table suggests the meaning of "one world". The stairway linking two floors uses jades at the bottom and auspicious clouds on both sides. At the inspiration of vast glass wall curtain, the designer turned the crystal pendants into the patterns of auspicious clouds and moon, filling the gap incurred by the hollow hall.

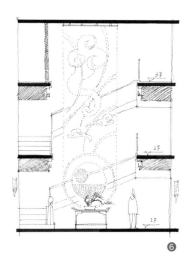

主卧是居者私密的休息空间，在视觉及陈设上，设计师选用深色调为软装配色，些许的配饰点缀还原空间的舒适感。以祥云图案装饰天花，地毯选用玉莲图案，以山水画为背景，透过落地窗向外望去，一线的珠江风景映入眼帘，美景与图案相映衬，设计师无不将生活的美好呈现于空间之中。开放式的衣帽间，方便了男女主人挑选衣服，亦是展示居者品味之所。

As a private rest space, the main bedroom applies a dark tone with some cozy furnishings. The auspicious clouds at the ceiling, the lotus on the carpet, and landscape painting at the background, echo the full river views through the French windows, unfolding the beauty of life. The open cloakroom on one hand makes it convenient for the master and mistress to choose clothes, on the other hand displays the owner's taste.

❶ 主卧设计手绘图。
The master room sketch.
❷ 独特衣帽间的设计满足女主人爱美的需求。
The unique cloakroom design satisfies the hostess's need for beauty.
❸❺ 设计师将美好的生活愿望转化为空间语言，并将其呈现在细节中。
The designer turns the good life wishes into space language and presents them in details.
❹ 顶层复式二层平面图。
The penthouse second floor plan.
❻ 主卧卫生间。
The master bathroom.

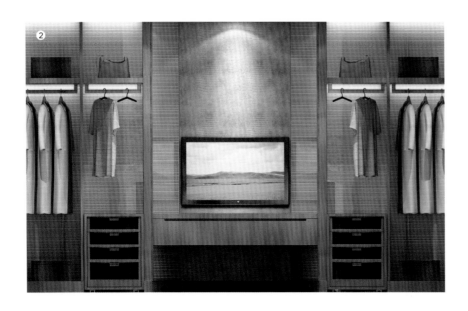

❶❷ 奢华的基调延绵整个空间，次卧的设计上，以繁花元素做点缀，将自然的景色带入室内，自然与室内空间在此一脉相通。
The tone of luxury fills in the whole space. As for the secondary bedroom, embellished with flower elements, it brings in natural landscapes, connecting the nature and the interior space.

❸ 在影音室中，居者可独享片刻的休憩。
In the video room, the residents can enjoy a moment's rest.
❹❺ 次卧卫生间。
Secondary bathroom.

 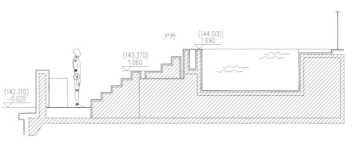

❶ 一线江景尽收眼底。
　A panoramic view of the river.
❷ 户外剖面图。
　Outdoor section.
❸❹ 居者可在此举行各式活动。
　Residents can hold various activities in the outdoor.
❺ 顶层会所户外平面图。
　The penthouse outdoor plan.
❻ 户外泳池。
　Outdoor swimming poor.

楼顶的空中花园，颇有"一览众山小"之观感。天际泳池与休闲洽谈空间结合的空中花园，满足了主人平日的休闲娱乐，亦可在此举行独立的派对，与亲朋好友坐落于此，如诗中所言"海上升明月，天涯共此时"。夜半星空，与家人在此览一城之景，享都市奢华人生。

The combination of rooftop swimming pool and drawing room at the hanging garden satisfies the owner's need of daily leisure. People can invite friends over for parties or they can just spend time with the family here at night, enjoying the luxurious urban life.

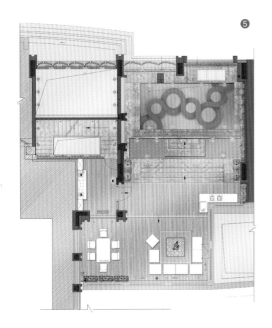

南京富力湿地会所
CLUB OF R&F WETLAND PARK, NANJING

关键词：蓝色、华贵、大气
Keywords: blue, luxury, magnificence

建筑设计 / 堂杰国际设计
景观设计 / 美国SWA景观设计事务所
建筑面积 / 58,000 平方米
委托范围 / 硬装及软装设计
委托面积 / 871 平方米

富力十号，临340万平中央公园，聚600年运粮河蜿蜒水系的婉约风情，毗邻繁华城市CBD之地，将经典与浪漫演绎到南京，呈现出中西交融的华贵与浪漫。

我们尝试将"移步换景""框景""对景"等东方园林的设计与西方的ART DECO糅合，巧妙的家具搭配、讲究的色彩纹样、精致的制作工艺，在新风潮中找到了中西合璧的交汇点。整体设计简洁之中不乏细腻的雕琢，在沉稳大气的整体构造中，糅合法式廊柱、雕花、线条等细节设计，外观华丽，彰显灵魂之优雅。

R&F No.10, located in the central park with 3.4 million square meters, is endowed with graceful and composed style from the 600-year-old canal. Next to CBD, it brings classic and romance to Nanjing, presenting the luxury and romance of eastern and western culture.

We tried to combine the unique designs of oriental gardens and the gentleness of the western ART DECO. The houses feature in clever furniture matching, exquisite color pattern and delicate craftsmanship with the intersection of two cultures found in the new fashion trend. The overall design is simple but with exquisite carvings. And the French columns, carvings, lines and other details are blended to present a gorgeous appearance, highlighting the elegance of the soul.

❶❹ 设计师将此打造为典雅的私人会所。
The designer made this a graceful private club.
❷ 会所二层平面图。
Second floor plan.
❸ 会所一层平面图。
First floor plan.
❺❻ 空间中的软装配饰。
Decoration in space.

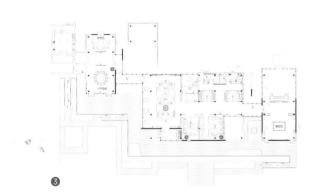

接待前厅，设计师以大理石柱体架构空间，加入金属线条点缀，呈现奢华高贵的视觉效果。走道两旁放置了艺术雕塑和挂画，阳光透过落地玻璃照亮室内空间。水吧区，巨型的水晶吊坠夺人眼球。公共洽谈区以静谧的蓝色与金色相映衬，线条与花纹的交错，打造艺术空间气氛。VIP洽谈室设置在一旁，以灰色与棕色为主色，打造高雅舒适的私人会所。

In the reception hall, the designer uses marble columns and metal lines to present a luxurious and noble view. Art sculptures and paintings are placed on both sides of the aisle while the sunlight illuminates the room through the French window. The huge crystal chandelier is eye-catching in the bar. The public negotiation area creates an artistic atmosphere with blue and gold, lines and patterns interlaced with each other. The gray and brown VIP negotiation room is set aside as an elegant and comfortable private club.

❶ 设计师利用楼高的优势打造两层空间，让空间视觉变大。
The designer uses the advantage of building height to build two layers of space to make space vision bigger.
❷ 二层餐厅。
Second floor's restaurant.
❸❹ 洽谈区。
Negotiation area.
❺ 组合吊灯。
Assemble lamp.
❻ 会所接待前台。
The reception of the club.

二层空间延续首层金色的奢华基调，设计师选用木质材料与之搭配，木色的温润弱化了奢华。"江南佳丽地，金陵帝王州。"江南的美景与南京城的繁华，都化作空间语言展现于空间当中，打造金陵之地的低调奢华场所。

The second floor continues the luxurious gold tone of the first floor. The designer chooses wood material to match it as the mildness of the wood color softens the luxury. The beauty of the south of the Yangtze River and the prosperity of Nanjing City have been displayed in the house, creating a modest luxurious place in Nanjing.

淮安金奥国际销售中心
SALES CENTER OF KINGTOWN INTERNATIONAL CENTER, HUAIAN

关键词：中式、简约清晰、动静分明
Keywords: Chinese style, simple and clear, distinct division of function areas

建筑及景观设计 / B+H设计师事务所
建筑面积 / 710,000 平方米
委托范围 / 硬装及软装设计
委托面积 / 3,400 平方米

简单的手法往往能表达纯粹的思想，卸下枷锁回到设计师的初衷。本案无论是硬装的隔断还是软装的家具都是棱角分明的直线条，加之沉着冷静的色彩搭配，表达出设计师成熟大胆的设计思路。中式配饰的引入，是设计师对新东方风格的一次大胆尝试，亮黄色的挂画和灯饰，无疑给整个空气带来别样生机。

Simple techniques often express pure thoughts. This case features in straight lines giving it a hard-fit partition or a soft-packed furniture. Combined with a composed and calm color scheme, it expresses the designer's mature and bold design ideas. The introduction of Chinese accessories is a bold attempt by the designer at the new oriental style. And bright yellow paintings and lightings undoubtedly bring a different kind of vitality to the whole space.

接待台 RECEPTION

入口与售楼部的大堂由一条艺术长廊连接，大有曲径通幽之感。走廊的尽端是一个宽阔的影视厅，右转即是大堂的入口，大有豁然开朗的感觉，这也得益于设计师先抑后扬的手法。售楼部的大堂是整个项目设计的核心，一个13米长的吧台将整个空间一分为二，既为客户提供小酌畅聊的小天地，又将其与沙盘、洽谈区隔开，空间分布简约清晰、动静分明，最大化地保护客户的私密性。

The entrance and the lobby of the sales department are connected by an art corridor, at the end of which is a movie theater and turning right is the entrance to the lobby. The lobby is the core of the whole project. A 13-meter-long bar divides it into two, on one hand providing a small area for customers to drink and talk with each other, on the other hand separating the area from the sand table and the negotiation area. The distribution is simple and clear with distinct division of function areas, maximizing the privacy of the client.

❶❷ 携手岭南画派艺术家,为项目定制折叠山水画屏风气势恢宏。
The folding screen of magnificent landscape painting is customized for the project together with Lingnan-style artists.

❸ 从走道的里面图可见随处的中式点缀。
From the elevation of the aisle, we can see Chinese embellishment everywhere.

❹❺ 设计师通过以小见大的手法,将山、水、竹等传统中式意象,藏于装饰之中,既保持空间的简洁,又保留了新中式的美学元素。
Traditional Chinese images including mountains, water and bamboo are hidden in the decoration by means of gimmick. By doing so, it maintains the simplicity of the space and preserves the aesthetic elements of new Chinese style.

❻ 水吧区。
Water bar.

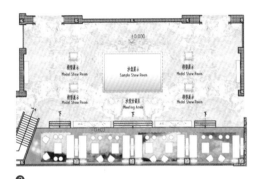

洽谈区坐落处以落地的玻璃墙为景，在此商洽交谈，静享自然光景，中式风情。顺着楼梯上至二层，VIP室与茶室的装饰简约高雅，现代家居与中式家居，在空间中完美的融合，些许中式元素的点缀，让空间更加统一和谐。

The negotiation area is against a French window, where you can talk and appreciate the natural views as well as the charm of Chinese culture. Going along the stairs to the second floor, the VIP room and the tea room are decorated in a simple and elegant style. The modern and Chinese furnishings are perfectly blended in the building, and some Chinese elements are applied to make the space more unified and harmonious.

❶ 中国的传统建筑设计注重内外互通。设计师将售楼部打造成"玻璃盒子",室外风光亦可在室内一览无余。
China's traditional architectural design pays attention to internal and external interoperability. The designer made the sales department a "glass box", and the outdoor scenery can be seen in the room.

❷ 销售中心平面图。
The sale center plan.

❸ 设计师通过巧妙的灯光布置,解决了空间中近十米长的高挑玻璃幕墙所带来的光照冲突。
The light conflict caused by the tall glass curtain wall of nearly ten meters in the space is solved by the designer with the subtle arrangement of lighting.

❹❺❻ 极具建筑体态的折转楼梯看似"悬浮"于空间之中,实际上它的角度、跨度及称重均在设计师的把握之中。
Spiral stairs featuring architectural shape are seemingly "suspended" in the space. In fact, the angle, span and weighing are completely controlled by the designer.

❶❷ 些许花艺的点缀让空间不失单调。
A bit of floral ornament lets a space do not break drab.
❸ 书吧空间分布图。
Space distribution of Reading area.
❹❺ 设计师注重空间对称性与层次感。
Designers focus on space symmetry and hierarchy.
❻❼ 二层餐厅效果图。
Second floor restaurant renderings.
❽ 将中式的元素放置于空间当中，是设计师对新东方风格的一次尝试。
To put the Chinese elements in space, it's a designer's attempt to make a new Oriental style.

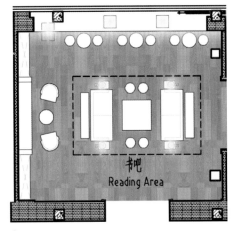

竹林、石狮、屏风、文房四宝……无处不在的中式元素，恰似一篇关于东方茶道的故事娓娓道来。现代建筑的体量感在这些元素的装饰下，更增了人文的感官享受。禅意、清幽、自然——现代建筑原本的冰冷和距离感就此被巧妙的设计淡化了。

Bamboo forest, stone lion, screen, and four treasures of the study……These Chinese elements are like a story about the oriental tea ceremony. The modern architecture, with the decoration of these elements, increases cultural pleasures. Zen, quietness, and nature -- the clever design has softened the original sense of coldness and distance in modern architecture.

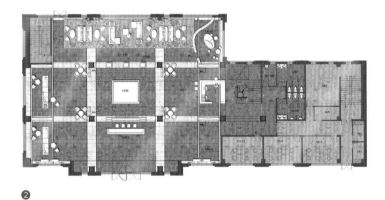

① ⑤ 销售中心沙盘区。
Sales center's sand area.
② 销售中心平面图。
Sale center floor plan.
③ 销售中心水吧区。
Sales center's water bar.
④ 阳光透过玻璃洒满整个空间。
Sunlight streams through the glass and fills the space.

走入室内，沙盘区盘旋的吊灯占据整体视觉空间，一旁些许陈列展柜为隔断与装饰，间隔着沙盘区与洽谈区。洽谈区以白色的现代家具为主，中式的家具做陪衬，绿植为点缀，用简单的装饰语言，还原空间。散座区，以黑色皮椅与白色大理石桌子装点空间，黑白搭配，干练简约。

Walking into the room, one first spots the hanging chandelier above the sand table. Between the sand table and the negotiation area stand some showcases for partition and decoration. The negotiation area features in white modern furniture accompanied by some Chinese furniture and green plants. Extra seat area applies black leather chairs and white marble desks- simple and neat.

① 楼梯的连接上下两层空间，如同连接古代与现代的中式风格的对话。
The connection between the stairs and the two floors is like a chinese-style conversation connected to the ancient and modern.
② 销售中心洽谈区。
Sale center's negotiation area.
③④ 花艺的点缀让空间充满生机。
The decoration of flower art makes the space full of vitality.
⑤ 坐落于此，与阳光邂逅。
Situated here, encounters with sunshine.

装饰柜体相间着洽谈区与水吧，木质的天花也延用于此，白色的花瓶造型吊灯与白色皮椅相配。后置的金属吊架，摆放着各式的艺术装饰品，使得空间不会过分单调。透过金属吊架，视觉可通至楼梯，空间之间没有明显的间隔。楼梯处一株植物矗立在旁，植物与木质的墙体呼应自然之美。本案中，设计师打破传统的售楼部设计，以开阳的手法使得不同空间相通相连。配饰细节处用中式元素装点空间，没有过分点缀装饰，以简约的手法点缀空间，使得空间更加开阔明亮。

Showcases between the negotiation area and bar area continues to use wooden ceiling and white vase-shape chandelier echoes the white leather chairs. The metal hanging shelves behind display all kinds of ornaments, making the room less monotonous. Through the shelves sees directly the stairway. A plant beside the stairway gives the room vitality and echoes the wooden wall on the beauty of nature. The designer breaks the tradition on sales department designs and connects different function rooms in an open structure. Furnishings use Chinese elements without redundant ornaments in a simple manner, making the room more bright and spacious.

❶❷❹ 思考建筑、景观与室内的关系，并在空间内部实现三者最佳的融合方式。
The best integration of architecture, landscape and interior should be achieved within the space on the basis of considering their relationships.
❸ 销售中心平面图。
Sale center plan.
❺ 销售中心沙盘区。
Sale center's sand table area.

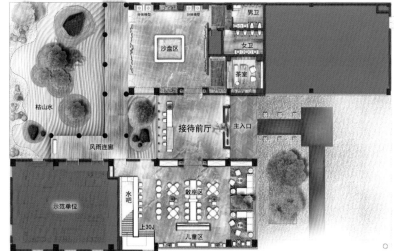

沙盘区选用大面积的木饰面材料，沙盘处于中心，一旁是LED电视，另一侧则是屏风隔断，远处的装饰物在灯光下尤为耀眼。卡座区是空间中独立的区域，与一旁的散座区无明显的间隔。两面落地的玻璃，朝阳的余光透过玻璃点亮了室内的空间。背靠一面书柜，在此可以书会友，千般浮华，莫若一纸书香。

The sand table in wooden veneer is in the center of the room with LED TV on one side and screen for partition on the other. The booth area is independent with no particular partition from the extra seat area. Sunlight illuminates the room through the French windows on both sides. On one side of the wall stands book shelves where one can meet friends and enjoy the pleasure of books.

❶❷❸ 俱木结构斜屋顶的设计与散落在空间各处的中式配饰，营造古典的中式风。
❹❺❻ All-wood and pitched roof, together with the pervading Chinese furnishings, creates a classic Chinese style.

❶

❷

❶❷ 为了满足空间的功能使用，设计师根据视觉及动线，对洽谈区做了合理的布局。
In order to meet the functional use of space, the designer according to the visual and dynamic line, the negotiation area made a reasonable layout.

❸ 水吧隐秘于楼梯中。
The water bar is hidden in the stairs.

❹ 洽谈区一角。
A corner of negotiation area.

❺ 设计师在洽谈区设立书柜，丰富了空间功能。
The designer set up bookcase in the negotiation area to enrich the space function.

散座区的装饰延用木饰面材料，软装上以金丝楠木间隙着固装的分隔中，沉稳的木色穿插在软装家具里，更具稳重感。看似繁复的分布其实是设计师为来访者营造轻松的气氛，淡化销售的功能，来访者坐落于此不经意间忘却了此处竟是销售中心。

Extra seat area continues to use wooden veneer. For the soft-packed furniture, phoebe zhennan is applied to separate the hard-fit partitions, the wood color of which gives an air of steadiness. The seemingly complex furnishings are actually meant by the designer for a relaxing atmosphere and reducing the sales function so that visitors would forget they are in the sales department.

水吧隐秘于楼梯回旋处,白色的大理石与浅木搭配成别样的风景线。设计师将原本的楼梯做了延伸,让楼梯看起来修长。从二层往下看,虽然看似空无,却呈现了空间多维度的延展。一联书法足以让空间充斥着文化气息。茶室的色调更浅一些,让来访者能在空间中静品一抹茶。

素雅的木色与绿植搭配贯穿整个销售中心,室内室外的交互,令室内空间充满生机,轻软装,去售楼化的设计,轻松安逸的环境还以都市人一丝舒适,能在这静谧庭院中,品味书香,以茶会友。

① 延展的楼梯。
  Extended staircase.
② 水吧区效果图。
  Effect drawing of water bar.
③ 走廊上的挂画与雕塑。
  Hanging paintings and sculptures on the corridor.
④ 茶室效果图。
  Effect drawing of tea room.
⑤ 二层平面图。
  Second floor plan.

The bar hides under the corner of the stairway. Its white marble and light-color wood create a nice view. The designer extends the original stairway and makes it appear more slender. Looking down from the second floor, one might see nothing but find out multiple dimensions. A couplet of Chinese calligraphy fills the room with cultural ambience. The tone of the tea area is even lighter, allowing visitors to rest in a more serene space.

The simple but elegant wood color and green plants fill in the whole sales department. The interaction between interior and exterior landscapes gives the room vitality. The design features in "being light", "de-marketing", and "relaxing and cozy space" gives urbanites a piece of comfort where they can read books, drink tea, and enjoy the beauty of life.

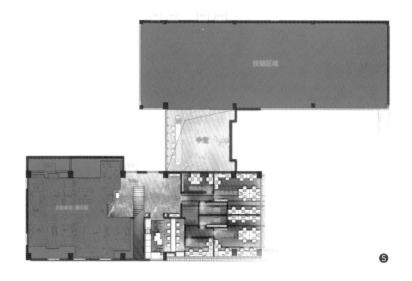

品牌开发商不断推陈出新，地产项目产品各有各的特点和诉求。我们不仅仅注重设计本身，也学会站在地产营销的角度反观设计，以适应市场的不断发展与变化。

Brand developers continue to innovate as real estate projects have their own features and demands. We not only focus on the design itself, but also actively learn to stand from the perspective of real estate marketing to adapt to the continuous development and changes of the market.

太原保利西江月
福州保利天悦
广州龙璟山
北京保利和光尘樾
厦门保利叁仟栋
Poly River Coast, Taiyuan
Poly Grand Mansion, Fuzhou
King's Mountain, Guangzhou
Poly Palace Of Light, Beijing
Poly Costal Mansion, Tongan

太原保利西江月
POLY RIVER COAST, TAIYUAN

关键词：新中式、盛唐、会所
Keywords: new Chinese style, Tang Dynasty, clubs

建筑设计 / 筑博设计股份有限公司
景观设计 / 广州怡境景观设计有限公司
建筑面积 / 304,009.11 平方米
委托范围 / 硬装及软装设计
委托面积 / 售楼中心 2,724 平方米
　　　　　样板A户型 180 平方米
　　　　　样板B户型 160 平方米

秦晋中原之地保存着中国传统人文历史的深厚与深情。借太原打通了时空的界限，回归到久远的古典世界去，静听盛唐故事。

太原，古称晋阳，与长安、洛阳并称李唐三都，为开国之君李渊起兵之地。项目坐落的晋源区，正是晋阳古城的所在地。西江月，原唐代教坊曲名字，后用为词牌名，名人雅士借此创作了大量的佳作。西江之上的泠泠明月照耀千古。从备受荣光的历史名城，到今日的现代化都市，城市面貌在时间长河中流变。古韵与新声的融合总是令人遐想，也为我们带来全新的设计思考。

Shanxi, the central plains during the rule of Qin and Jin, hosted Chinese traditional culture and history. With the help of the city Taiyuan, we break the boundaries of time and space, return to the ancient classical world and listen to the story of the Tang Dynasty.

Taiyuan is called Jinyang in the ancient time, which is one of the three capitals (the other two are Chang'an and Luoyang) of the Tang Dynasty. It is also the place where Li Yuan, the founder of the Tang Dynasty raised his army. The Jinyuan district where the project is located, is the location of the ancient city Jinyang. The clear moon above the Xijiang River shines through the ages. Its urban appearance has been changed over time from a historic city to today's modern city. Endless imagination could be sent together with brand-new design thinking thanks to the fusion of ancient rhyme and new music.

1

❶❷ 唐代建筑的木结构均不上漆。为了遵循传统，项目所用金丝楠木均保持了原漆的状态。
The wooden structures of the Tang Dynasty buildings are not painted. Thus, original paints of Phoebe sheareri used in the project are remained to follow the tradition.

❸❹ 祖母绿的点缀令人眼前一亮。
It's eye-catching with emerald embellishment.

❺❻ 中式花纹的窗格在墙壁投下的阴影，宛如一幅绝美的图画。
The shadow of the Chinese-style pattern pane cast on the wall is like a great picture.

❼ 首层平面图。
Plan of the first floor.

2

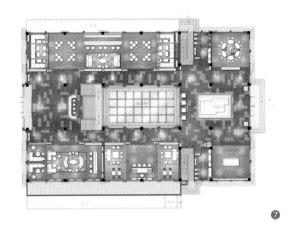

金丝楠木建筑的体态严整开朗且形体俊美。古典卯榫结构被珍贵的金丝楠木包裹，伫立在大堂的中轴地带，营造出巍峨醒目且金碧辉煌的华贵。地面使用深色的火烧面石材，风格粗粝凝练。充满中式元素的灯饰照亮庭廊，令人颇有几分指点江山的豪迈感。

The phoebe zhennan building features in neat and beautiful body. The classical tenon structure is wrapped in precious phoebe zhennan and stands in the middle of the lobby area, creating a kind of eye-catching and splendid luxury. Dark flamed stones are used on the ground with a bold and neat style. The lighting full of Chinese elements illuminates the gallery, giving people a sense of heroism.

正如《淮南子·说林训》所言："鹤寿千岁，以极其游。"设计师减少设计手法上的元素符号，汲取更深层的"鹤魂"，化为洽谈区的气质，体现晋中地带兼收并蓄的风格。灰金色调和火烧面的石材延续着沉稳大气的氛围。由屏风提炼出独特的花格元素，分别在落地窗、吧台和柜体的细节处，使其彰显高贵优雅和悠然深远的独特魅力。

The spirit of crane has been mentioned in *Huainanzi:Analects of Wood*. Designers reduce the symbolic elements of design techniques and draw deeper "crane-soul" into the temperament of the negotiation area. The gray gold tones and flamed stones remain the composed and calm atmosphere. The unique lattice elements inspired by the screen are shown in the details of the French windows, bars and cabinets, presenting the unique charm of elegance and leisure.

❶❷❸ 洽谈区的陈设用现代的优雅呼应华美的古韵，营造出非凡的气度。
Furnishings in the discussion area echoed the gorgeous ancient charm with modern elegance, creating an extraordinary temperament.

❹❺ 大量的灰石模仿古建筑中砖纹的拼接，中式古典花纹紧扣空间气质，尽显力量和厚重。
Slices of brick patterns in the ancient building are imitated with large numbers of limestone. Meanwhile, power and solemnness are shown with Chinese classical patterns being closely adapting to the space temperament.

❻❼ 唐代形成完备的学校教育制度，因此国学培训区模拟"二馆"的布置，设案几书架，营造浓厚的历史文化氛围。
A complete schooling system is formed in the Tang Dynasty. Therefore, traditional tables for eating and reading as well as bookshelves are set to simulate the arrangement of "Venue II" in the training area of traditional culture, so as to build an atmosphere mixing history and culture.

❽ 洽谈区立面图。
Elevation of discussion area.

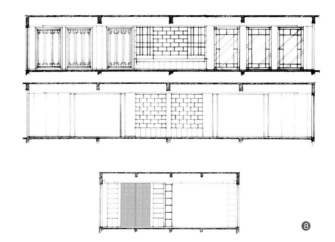

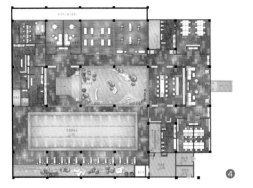

饮茶写字、读圣贤书自古以来都是中国人涵养性情的独特方式。茶艺区的案几端正，材质自然，让人或坐或立，品悟"道"和"理"。借用廊架细分会谈区域，既保证私密性之余且不妨碍景色的互通，强调专属中式对称、平衡之美。负一层为大型综合休闲区，具备娱乐、健身、餐饮等功能。在延续首层的设计手法之外，顺着中庭及侧庭的景观优势，采用大面积玻璃借景入室。早晚充足的光线带来丰富的视觉享受。

Drinking tea, writing calligraphy and reading books written by sages have been a unique way for Chinese people to develop their disposition since ancient times. At the upright tea table made of natural material in the tea art area, one can be seated or stand up and comprehend the "Dao" and "Li". The gallery is designed to subdivide the reception area, which not only guarantees the privacy without blocking the continuity of sceneries, but also emphasizes the beauty of Chinese symmetry and balance. The basemnet floor is a large-scale integrated recreation zone with entertainment, fitness, catering and other functions. In the continuation of the design tactics on the first floor, taking the landscape advantage of the atrium and the side court, a large scale of glass is installed to let sceneries come into view from the inside.

❶ 通往负层会所的通道以双梯回廊的形式呈现。
The access to the clubhouse in the negative floor is presented in double ladders.
❷❸ 延续别致的新中式风格的简餐区及茶艺区。
Quick meal area and tea ceremony area featuring new Chinese style are extended.
❹ 下沉式建筑引用中景以及边景的方式,打破以往地下层缺少光照以及景观的设计难点。
The medium and side shots are adopted in the sunken buildings, breaking the design difficulties of the lack of light and landscape in the negative layer in the past.
❺❻❼ 健身区外葱茏的绿意和自然的天光令人卸下烦嚣,大隐于市,回归心灵的平静。
The greenery and natural skylights outside the fitness area allow people to relax and return to the peace of mind through being implicit in the city.

现代手法 古典大美

Modern Techniques, Classical Beauty

本案将聚焦户型全生命周期,预设了多种家庭结构下的生活场景,在设计层面为空间长远使用做足功课。着重将空间功能进行重新的整合和预留。180m²户型具有大尺度的空间资源。原建筑的面宽相当优越,很好地适应北方秋冬时居住空间的采光保暖需要,品字形的户型结构利于南北对流。室内设计顺应户型特点,强调宽厅、宽墅的概念,提供高品质的舒适生活。

With a focus on the whole life cycle of an apartment, this proposal will presuppose multiple life settings of different family structures, and do the groundwork for the long-term use of space at the design level, emphasizing the re-integration and reservation of the space functions. The 180-square-meter apartment has large-scale space resources. With a perfect width, the original building can meet the lighting and warming needs of the living space in autumn and winter in northern China; also, the apartment structure in the shape of Chinese character pin is beneficial to north-south flow. The interior design should emphasize the concepts of wide hall and wide villa on the basis of following the apartment types, so as to provide a high quality and comfortable living.

❶ 中央走道是全屋动线的重要节点。
The central hallway is an important node of the motion line for the apartment.
❷❸❹ 餐厅到客厅的纵深将近10米,大尺度带来无与伦比的舒适感。
The depth from the dining room to the living room is close to 10 meters, which can bring unparalleled comfort.
❺ 厨房采用磨砂玻璃门作隔断,保持室内整体的明亮感。
A frosted glass door is adopted as a partition to separate the kitchen, maintaining the brightness indoors.
❻ 180m²户型平面图,空间复合功能在设计之初便考虑入内。
Plan of 180m² apartment, the composite function of space is considered in design at the beginning.

①

② ③

❶ 金属质感的茶几上摆放着卷轴，现代造型的沙发裹着中式的山水波纹，色调沉稳的直纹黄杨木以面块渲染气氛，现代设计手法与古典美学完美融合。
There are scrolls on the metal tea table. The modern sofa is wrapped in Chinese-style landscape ripples. The straight-lined boxwoods in calm tone create atmosphere in pieces. It's a perfect integration between the modern design methods and classical aesthetics.

❷❸ 书房从独立走向开放，体现着空间复合功能的提升。笔墨纸砚、瓷器名画无不流露着浓厚的文化气息。
The study is set openly from independently, indicating the improvement of composite functions of space. A strong cultural atmosphere is presented with writing brushes, ink sticks, paper and inkstones.

❹❺ 紫色寓意尊贵。卧室之间辅以不同饱和度的紫色以作区分，增加视觉上的层次性。
Purple means notability. Purples of different saturation level are adopted to distinguish bedrooms, increasing the visual hierarchy.

设计糅合中式古典和现代简约的手法。立面干净，色调优雅。大地色系各有层次，含蓄有格调。选用的家具造型简练，配饰则尽显东方浓厚的文化气息，呼应当地深厚的文化底蕴。

The residence incorporates Chinese classical and modern simple design techniques. The facade is clean while the tone is elegant. Earth colors have different shades, reflecting implicit styles. The style of the chosen furniture is sophisticate, and the accessories are full of oriental culture, echoing the local profound cultural heritage.

❶❷❸❹ 结构类似的空间可以通过细节形成截然不同的气质。色调更为明快简约的木饰面和更多的金属饰品令空间充满了现代的精致。
Completely distinct temperaments can be formed through details in spaces with similar structures. Modern dedications are filled in the space thanks to wood facing in bright and simple tones as well as many metal decorations.

❺❻❼ 明亮的厨房和餐厅充满居家气息。
The kitchen and dining room with brightness and living atmosphere.

❽ 160m²户型平面图。
Plan of the 160-square-meter apartment.

⑤
⑥ ⑦

160m²户型结构与上述户型大致相同，设计手法上则更采用时尚感更强的港式现代风格。以直纹白杨木为饰面的空间更为明快利落，映衬着太原现代化风貌。

Being about equal to the above apartment in structure, the 160-square-meter apartment is designed in a Hong Kong-style modern style featuring a strong sense of fashion With straight-lined box-woods straight as its facing, the clear-cut and bright space shows a modern style of Taiyuan.

⑧

❶❷ 主卧运用了不少透明亚克力、黄铜、几何抽象图案等视觉元素以延续整体的现代风格。
A variety of visual elements such as transparent acrylic, brass, and geometric abstract patterns are employed in the master bedroom to extend the modern style as a whole.

❸ 主卧卫生间。
The bathroom of the master bedroom.

❹❺ 次卧的功能属性可根据需求作调整。软装上沿用类似的设计思路，卧室辅以不同饱和度的蓝色作区分。
The functional of extra bedrooms can be adjusted as needed. A similar design idea is adopted in the soft decoration. And bedrooms are distinguished with blue of saturation levels.

主卧朝南，延续主色调的沉静典雅，卧室的配饰仍为简约、实用为主。次卧位于户型的西北侧，可以根据业主需要，调整为月嫂房、孩子房等，令空间的可变性大大提升。

The main bedroom faces south and boasts plenty of sunlight. Continuing the calm and elegance of the dominant tone, the bedroom accessories are still simple and practical. The secondary bedroom can be adjusted to the nanny's room, children's room and other rooms according to the needs of the owner, or it can be directly connected to the main bedroom so that the latter enjoys the advantage of running through the north and the south.

scape city. It's really a glory for designers to have designs full of thoughts and respects in Fuzhou, a national historical and cultural city with more than 2,000 years of history.

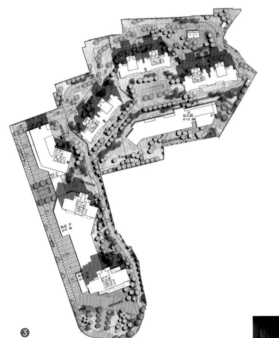

被誉为八闽首善之区的福州鼓楼区，为福州全市的经济、文化、政治中心及现代金融服务业中心，而福州保利天悦花园正位于此。建筑群以 L 型排布，尽享 80 公顷湖景公园的自然景观，吸纳屏山文化底蕴，系涵盖多种空间类型的高品质社区。水色生辉，既宜安居也宜乐业，人们可以在此找到自己的节奏，也找到属于自己的"安居之所"。

Gulou District of Fuzhou, known as "the best district in Fujian", is the economic, cultural, political and modern financial service center of Fuzhou. Fuzhou Poly Garden is located here, whose buildings are arranged in an "L" shape, enjoying the natural landscape of the 80-hectare lakeview park and absorbing the cultural heritage of Pingshan. It is a high-quality community covering a variety of space types, suitable for both living and working. People can find their own tempo and "paradise" here.

❶ 项目综合住宅和商业等业态，建筑外观具备现代主义风格。
Consisting of complex and commerce, the exterior of the building featured a modernist style.
❷❸ 项目规划总图。L 型格局确保每栋建筑均可观赏到西湖、屏山公园的景观。
General plan of the project planning. All buildings can enjoy the scenery of the West Lake and Pingshan Park in the L-shaped layout.
❹ 销售中心外的流水走廊在夜色中更显迷人。
The water corridor outside the sales center is charming at night.

铜立方里的水文精灵

湖心春雨，静聆风声。半透明的销售中心建筑在阳光下熠熠生辉，宛如一个精致且受人瞩目的盒子。从北京华尔道夫的"铜立方"汲取灵感，建筑的肌理立体舒缓。以活泼池水为引，客人信步走入室内，感受新亚洲装饰风格所带来的精妙光影和儒雅气息。

Elf in the copper cube

The translucent sales center shines in the sun, like an exquisite and eye-catching box. Inspired by the "Copper Cube" of Waldorf Astoria in Beijing, the texture of the building is stereoscopic and soothing. Following the pool water into the room, the clients can feel the delicacy and elegance of the new Asian decorative style.

1. 销售中心接待前台区域实景。
   The picture of reception areas in the sales center.
2. 大堂效果图。流云状的灯饰与大理石对比，相得益彰。
   Rendering of the lobby. The cloud-like lighting and the marble bring out the best in each other.
3. 销售中心平面图。
   Plan of the sales center.
4. 福州寿山的石雕艺术享誉海内外。石材的巧妙运用也是该项目的设计亮点之一。
   The stone carving art of Shoushan, Fuzhou is well-known at home and abroad. Using stones skilfully is one of the highlights of the project.

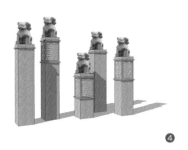
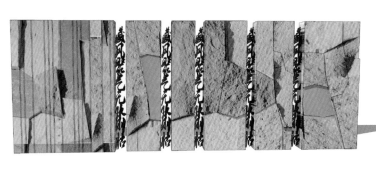

儒雅者安静沉稳，不轻易显山露水，然而一旦出手，也从不令人失望。从儒者的形象中汲取灵感，设计首先以坚实的大理石为空间气质奠定基调，横平竖直的主体线条构成了空间的主骨架。陈设装饰正如儒雅之人的质地，总是纯粹平和的。

Inspired from the image of the Confucian (composed and modest, but never fails others in action), the designer sets the tone with solid marble as the horizontal and vertical lines constitute the main framework of the building. Furnishings are like the texture of the Confucian, always pure and peaceful.

❶❷ 洽谈区效果图。隔断如屏风排列严谨，正面可置物，背面则悬挂插画。如此重复，充满了视觉上的纵深感。
Rendering of the discussion area. Partitions are rigorously arranged like screens, which is full of depth.

❸❹ 湖蓝色和橙色的碰撞新颖别致，为空间增添一丝新古典式的优雅。
Novel and unique collision between the lake blue and the orange is novel and unique gives the space a neo-classical elegance.

根据空间的结构，设计师把洽谈区统一布置在最佳的赏景区域，并点缀以鲜亮的橙红色作为点睛之笔，隔断巧妙地分隔开各个独立的会谈区，装点其上的梅画、书籍、折扇等装饰元素，呼应着儒雅者不落俗套的审美，将福州地区深厚又婉约灵动的文化蕴底展露无疑。

Based on the room structure, the designer endows the negotiation area with the best views and embellishes the bright orange-red as the finishing touch. The partitions subtly separate the independent meeting areas and are decorated with wintersweet paintings, books, folding fans and other elements, echoing the unique aesthetics of the Confucian, unveiling the profound and inspiring cultural heritage of Fuzhou.

❶❷ 基于精装交付标准，软装中，爱马仕式橙色十分亮眼，呈现出优雅精致ArtDeco风格。
Based on handover standards of delicately decorated apartments, Hermes orange is very eye-catching in the soft decoration, showing the elegant and refined ArtDeco style.
❸ 户型平面图。
Plans of apartment types.
❹❺❻❼ 客厅效果图及实景图。
Rendering and photo of the living room.

雅痞与现代的优雅对话

优雅户型面向谦逊又自信的精英人士，设计用一点点雅痞的灵动俏皮，搭配ARTDECO的轻奢，营造出活泼且富有品质的家居空间。以米白色为主调，辅以鲜亮的橙红色，极具冲击力的视觉色彩让时尚感油然而生，既呈现居者独特的品味，又满足营销端口的需要，令导览过程显得明快。

Conversation between yuppie and modern style

The elegant units, for the modest and confident elites, are designed with a little bit of playfulness of yuppies and combined with the luxury of ARTDECO to create a lively and high-quality space. With creamy white as the main tone, supplemented by bright orange-red, such colors with huge visual impacts overflow with fashion sense. It not only presents the unique taste of the owner, but also meets the needs of marketing, making the clients refreshing during the tour.

❶❷ 起居室中预留品茶区，为空间的不同功能使用做预留。
Tea area is reserved in the living room for the use of different functions of the space.

❸❹❺ 餐厅与厨房继续以直线条的理性为基础，辅以棕色点缀，营造更温馨的饮食空间。
Based on the rationality of the straight line, the restaurant and the kitchen are embellished by brown, creating a warm dining space.

❻ 户型平面图。
Plan of apartment.

现代户型中，家具的巧妙应用对空间气质起到至关重要的作用。品茶休憩区采用中岛型结构，与通高整排的书柜相融，茶香与书香天生一对。与客厅相连的餐厅，采用原木的间隔墙光暗交错，优雅律动感油然而生。简约自然的玻璃吊灯与整体色调风格呼应，与黑白色的餐椅上泼洒出怡人的暖意。

For modern units, the clever application of furniture plays a vital role in space temperament. The tea area applies island-shaped structure, complementing the bookcases, where the tea fragrance and book fragrance make a perfect fit. The restaurant connected to the living room uses wooden partition, creating a sense of elegance and rhythm. The simple and natural glass chandelier echoes the overall tone and casts warmth on the black and white dining chairs.

经典的棕色系，在巧妙的灯光点缀下，晕染出类似焦糖的甜蜜感和踏实感。细节处的繁华与葱郁的绿意，装饰与简约的平衡，令人遐想。

The classical brown style of ArtDeco design, with subtle lighting, gives a caramel-like sweetness and security. The balance between decoration and simplicity as wells as the imaginative jazz-era style blended with tea, reinterpret the "classic" spirit.

❶❷ 主卧的设计仍然讲求简洁。干净的立面，点到即止的装饰，使整体呈现出舒服宁静的视觉效果。
The master bedroom is designed in simplicity. Clean facade and reserved decoration present a comfortable and quite visual effect.

❸❹❺ 步入式衣帽间精致实用，与卫生间相连，十分方便。
Exquisite and practical walk-in cloakroom is connected to the bathroom, which is very convenient.

❻❼ 长辈房与儿童房。
Elders room and children's room.

大都会的秩序：理性与感性的融合统一

时髦、个性、张扬、炫目，大都会将秩序和奢华糅合于一体。冷冽炫目的金属点缀，极简的黑白灰三调大界面，切割边缘明显的厚重石材，横平竖直的秩序感和高挑的层高诉说着现代设计手法理性与感性之间的融合与重构，令接待大堂散发着庄严恢弘的气度。

洽谈区的陈设中，褐色的木质茶几搭配金属支架、造型夸张的金属落地灯以及细腻触感的皮质。不规则的"云朵"造型灯，则将空间上下的轻盈与厚重做出了平衡。庄严的使命感，保证了私密空间的包裹性，其中不经意间跳出的亮色调打破了整体空间的沉闷。

❶ 利用建筑空间的围拢式结构布局洽谈区，冷冽炫目的金属线条营造出庄严的殿堂。
The discussion area is arranged in a closed structure using the building spaces; while a solemn hall is built with the cool and dazzling metal lines.
❷❸ 无处不在的金色元素强调着尊贵感。
A prestigious sense is highlighted by pervasive golden elements.
❹ 步入式的接待厅宏伟有序，设计师按照参观动线进行布局。
The grand and orderly walk-in reception hall is arranged in accordance with the motion line of visiting.

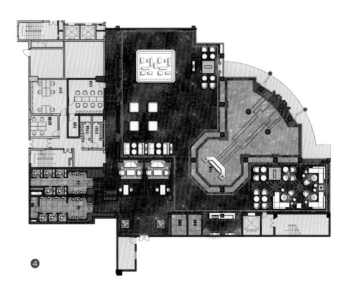

❶❷ 流光溢彩的酒吧区。
The glamorous bar area.

❸❹ 除了金色，偶尔出现的棕红、灰黑等又留下了丝丝神秘感。
In addition to gold, occasionally-appearing brown-red, and gray-black, etc. show a sense of mystery.

Metropolitan society: sense and sensibility

Fashion, personality, publicity, dazzle... The metropolis combines order and luxury. Cool and dazzling metal embellishment, three simple tones of black, white and gray, thick stones with distinct cutting-edges, in-order horizontals and verticals, high-rise buildings... they all express the integration and reconstruction between sense and sensibility of modern design techniques, which makes the reception hall solemn and magnificent.

For furnishings of the negotiation area, the brown wooden tea table is equipped with metal brackets, exaggerated metal floor lamps and delicate leather. Irregular "cloud" lights balance the lightness and weight of the space. The solemn sense of mission guarantees the inclusiveness of the private area, and the bright color that inadvertently jumps out breaks the dullness of the whole space.

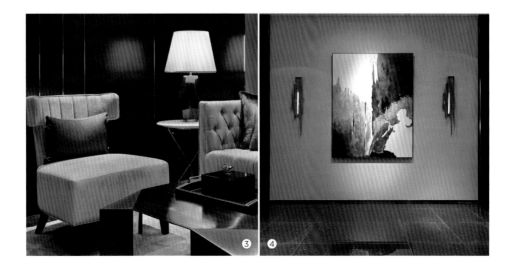

❶ 继承港式设计的轻奢风格，经典油画和法式的天鹅绒面料增添别样的浪漫。
Different kind of romance is added with classic oil painting and French velvet fabric on the basis of inheriting the slight luxury style of Hong Kong-style design.
❷ 户型平面图。
Plan of apartment.
❸❹ 平层户型设计讲究空间动线的合理性，绅士画像表露出主人对生活品位的追求。
The gentleman portrait shows the owner's pursuit of lifestyle. The design of flat floor plan emphasizes the rationality of the motion line of the space.

轻奢·生活·设计

双阳台分别对应客厅及餐厅，东、南、西面三面采光尽显大平层住宅的优势。代表稳定的米色贯穿于素雅的起居室中，空间整体引用了典雅的金色与线条勾勒的美感，使之结合带出直观的体感。多重石材与木饰为空间增添了更多沉稳舒适的感觉。

Luxury·life·design

A reasonable flow in a large flat building is especially important. With beige representing stability running through the simple and elegant living room, the space as a whole integrates exquisite gold and lines to make a visualized look. Multiple layers of stone and wood bring more calmness and comfort to the space. And the jazz portrait on the wall reveals the owner's pursuit for quality life.

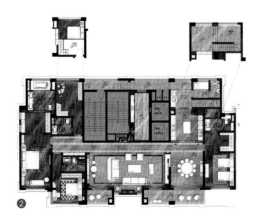

"生活就是需要仪式感,否则哪怕用山珍海馐来充饥饱肚,也是食之无味"。在餐厅的设计中,天花吊灯极具立体感,在银紫色餐椅和褐色金属圆桌的搭配下,空间的奢华之美尽显眼前,厨房的设置集中西方烹饪习惯,洗-切-煮-盛的备菜过程通行无碍,并在宽敞的厨房中增设岛型操作台,可以容纳多位家人同时入厨,一起享受烹调的乐趣。

In the design of the dining room, with the extremely stereoscopic chandelier, silver-purple dining chairs and brown metal round table, the luxurious beauty of the space comes into sight. The kitchen setting caters to both Chinese and western cooking habits with an unobstructed flow. And an island-shaped operating floor is installed in the spacious kitchen, allowing family members to enjoy cooking together.

❶❹ 宽阔舒适的餐厅。
A wide and comfortable dinning room.
❷❸ 中岛式流理台是港式设计的显著特点。
The island range sink has remarkable features of the Hong Kong-style design.
❺❻ 餐厅精致的配饰。
Exquisite decorations in the dinning room.

卧室巧妙地选用了宝蓝色、金色和银灰色搭配,点缀以颇具现代特点的艺术画作,稳重的气息和雅致的魅力展露无遗。

The bedroom, cleverly decorated with royal blue, gold and silver grey, embellished with artistic paintings with modern features, delivers an air of steadiness and elegance.

❶ 主卧彰显着浪漫华贵的精致主义。
Romantic and luxury refinement is manifested in the master bedroom.
❷❸❹ 功能齐全的主人套房。
Full-featured master suite.
❺❻ 水晶台灯和抽象的挂画成了艺术感的来源。
The crystal table lamp and abstract paintings are sources of space art.
❼❽ 客房的配置也丝毫不将就。
Extra rooms are also decorated perfectly.

郦道元《水经注》记载北京："左环沧海，右拥太行，北枕居庸，南襟河济。"深厚的历史底蕴和丰富的人文资源令北京既出落得大气雍容，又有接地气的人间烟火。和光尘樾项目坐落于北京 CBD 东扩路线之上。繁华高端的望京商圈、798 艺术区触手可及、国际商务中心以及第四使馆区令它拥有浓厚的国际化文化艺术气息。其次，项目兼有 30 分钟即达首都机场的完善周边配套，成就高品质生活社区。

Li Daoyuan recorded Beijing as: "It's surrounded by the sea on the left and hold by the Taixing Mountain on the right with Jurong Mountain in the north as well as Yellow River and Jishui in the south." on the "Shui Jing Zhu (Notes on Book of Water)". Beijing is both elegant and earthy thanks to its profound historical heritage and rich cultural resources. Heguang Chenyue Project is located on the eastward expansion of the CBD in Beijing. The prosperous high-end Wangjing business district, and the 798 Art District are near at reach. Meanwhile, it's also filled with a strong atmosphere of international culture and art with the International Business Center and the Fourth Embassy District nearby. Secondly, the project has a perfect surrounding facility of reaching the capital airport, becoming a high-quality living community.

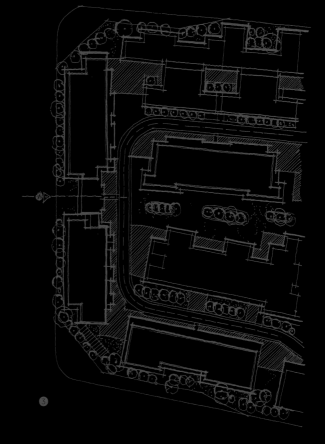

归途和等候亦备受关怀

SNP积极介入建筑项目的早期设计，为地下车库规划出科学的车辆行动路线，优化了项目楼宇车库大堂的平面布局，并以"互动"为题进行室内设计，务求让住户能在归家的路途中感受到一丝的暖意。地下大堂被棕色与灰色温柔包裹，大理石材质平整厚重，带出一种安宁和稳重，柔和的灯光设计，巧妙地中和了地下空间原本的冷峻感。

Cared when returning or waiting

We actively engaged in the early design of the project, planned a scientific vehicle action route for the underground garage, optimized the layout of the garage hall, and conducted interior design with the theme of "interaction", so that residents can feel cared for when returning home. The underground lobby applies brown and grey as the main tone, coupled with flat and thick marbles, delivering an air of calmness and steadiness. Soft lighting design cleverly neutralizes the original sense of coldness in the underground. Walking into the lobby.

❶❷❸ 项目定位为国际化生活社区，因此整体规划十分注重居住科学合理和周边环境氛围。
Since the project is positioned as an international living community, great attentions have been paid to the scientific and reasonable living environment and the atmosphere of the surrounding environment.

❹❺ 因应不同的产品单元调整地下大堂的装饰以贴合住户属性，达成差异化、个性化的内景观设计。
The decoration of the underground lobby is adjusted to fit resident attributes based on different product units, achieving a differentiated and personalized interior landscape design.

❻❼ 不断轮播天气信息的电子屏幕，富有动感的雕塑和雅致的枯山水微型景观，业主和宾客在等待中感受到无处不在的生活艺术气息。
An electronic screen constantly showing weather information, dynamic sculptures and elegant miniature landscapes of Japanese rock garden allow owners and guests to feel the pervasive art of living.

❶❷❸ 项目的建筑外观拥有典型的现代主义风格，体现出干净和理性的精神内核。整体线条简洁有力，采光面宽大胆且利落，将光影的艺术发挥得淋漓尽致。
The architectural appearance of the project has a typical modernist style, reflecting a clean and rational spirit core. The overall outline is simple and powerful; and the art of light and shadow is put to full use with the bold and neat face white of lighting surface.

❹ 由景观大师李宝章先生创作的景观，细节处见心思。
From details, you can see times and efforts spent on the landscape created by Mr. Li Baozhang.

光的演变，意味着时间的来去。以归心居所的理念解读当代的设计平衡之术，超脱单一的室内设计思维，更宏观地在建筑、在景观的角度上，一切以真实需求为依归，呈现居者内心深处的诉求。开扬通透的景观、动线合理的空间布局和强调交流的氛围，光景流转之处，即是优质人居所在。

The evolution of light means the passage of time. Interpreting the modern balance design with the concept of home-based residence, the designer can surpass the monotonous interior design and cater to the authentic pursuits of residents from the macroscopic perspective of architecture and landscape. Open and transparent landscapes, reasonable layouts of flows and focus on interaction... all of these make a quality residence.

❶❷ 户外的景观资源如同一幅优美的图画,赋予室内设计更多的灵感闪光。
Outdoor landscape resources are just like a beautiful picture, giving inspiration to the interior design.

❸❹❺❻ 敛起游走在现代社会的所有锋芒和锐气,何妨静观光的四时流转,天际线处的云卷云舒。
Lay all shines and spirits in the modern society aside and observe time to goes by in four seasons and clouds scud across at the skyline.

❼ 图中分别对应书房模式及会客模式。这是我们深入研究户型特性后提出的创新性处理手法。通过可移动的构件,复合不同的生活场景,令空间百变。
The figures correspond to the study mode and the reception mode. It's an innovative approach we proposed after going into the characteristics of apartments. Different life scenes are combined to enable the space changeable through movable components.

塞北园林,景观长廊

厂型的布局呈现出开放之态,又保留了围合之感。书房、客厅及餐厅有机地交融于一体。横厅具有高效的空间复合能力,可通过屏风的开合及家具的变化,灵活切换"书房模式"及"会客模式",以对应不同的场景需求。通高落地玻璃之外的公园景观随时间流逝而变化,从夏日的水木琳琅,到冬季的枯山寂水,一如江南园景。户外无限风光得以在这零隔断的景观长廊里呼吸,无间断地簇拥至人们的眼前。

Northern garden, landscape corridor

The L-shaped layout is open while retains the sense of envelopment. With full interaction among the study, living room and dining room, the three are integrated as one. The outskirts of the capital city have no boundless desert as the north of the Great Wall but creates unique scenery in midsummer or golden autumn. The scenery of the park through the floor-to-ceiling windows changes with the passage of time, from flowing waters and luxuriant trees in the summer to still waters and bare mountains in the winter, just like the garden scenery in Jiangnan. In this way, the outdoor scenery goes on and on in the full-view landscape corridor.

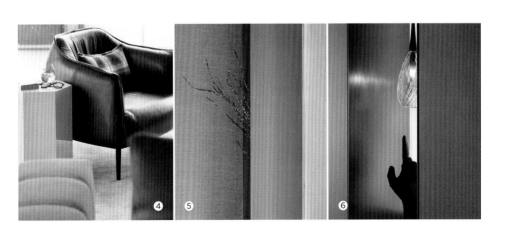

艺术思维也促使居家陈列逐新趣异,面对不同场合也应对自如。看似随意放置插画家Daria Petrilli的作品,映衬着空间的意式韵味。同样来自意大利的家具品牌Minotti亦可自由灵活地组合,以响应客厅不同情景场合的来回切换。细腻的品质细节,简洁的造型为客厅更添精彩。

Artistic thinking helps home-based furnishings to be new and interesting, and fit different settings. The work of the illustrator Daria Petrilli, which seems to be randomly placed at home, reflects the Italian flavor of the place. Furniture from the Italian brand Minotti can be freely and flexibly combined to fit different settings in the living room. Exquisite quality details and simple style make the living room more elegant.

❶ 饶有趣味的插画成为客厅艺术氛围的点睛之笔。
Interesting illustrations are a finishing touch of art atmosphere in the living room.

❷ 书房、客厅、餐厅在L型端厅布局中恰如其分地融为一体。
The study, the living room, and the dinning room are properly integrated as one in the layout of L-hall.

❸ 造型简洁的书柜和舒适的沙发是阅读体验的最佳搭配。
Compact bookcases and comfortable sofas are perfectly matched for the reading experience.

匠心所造，自见真章

"阳光厨房"，剔除东面的墙体，改由落地玻璃覆盖，白昼时分的厨房尽受自然光线的照拂，既环保又健康。"取洗切煮盛"的流程操作无碍，专业高效的厨房空间，让柴米油盐和人间烟火也因此有了实在的味道。

Ingenuity and genuineness

The "Sunshine Kitchen" removes the wall from the east and is covered by floor-to-ceiling glass. The kitchen at daytime is bathed in natural light, environmentally friendly and healthy. With unobstructed cooking flow, the professional and efficient kitchen gives daily cooking quite a hand.

❶❷ 凭借实践经验和针对性的推敲演绎，我们为该户型制定出一套高质量的全精装交付标准。
A set of high-quality whole-bound handover standards has been developed for the type of apartment with our practical experience and targeted deduction.

❸❹ 玄关既有置物、遮挡的功能，同时通过面板的开合参与到横厅情景模式的切换环节中，功能强大。
The hall way can be used for placing objects and occluding, which can also be a part of switching scene modes in the horizontal hall via the opening and closing of the panel with powerful functions.

❺ 通透的玻璃代替封闭的墙体，自然光的透入增加了室内的亮度，还带来了安全洁净的感觉。
Transparent glasses are adopted instead of the closed wall to allow the penetration of natural light, not only increasing the room brightness, but also bringing the safe and clean feeling.

❻❼❽ 厨房与餐厅、客厅没有隔断，不同活动区域的家庭成员之间可保持视线的交互。
The kitchen and the dining room, the living room are not separated, so that family members in different activity areas can keep eye contacts.

① ③ 卧室延续现代简洁的诉求，注重舒适性，我们通过舒缓的灯光氛围和柔软细腻、纹路经典的面料强调休憩时所需的宁静和体贴。
The modern and simple style is extended to bedrooms with a focus of comfort. Moreover, the tranquility and thoughtfulness for resting with a comfortable lighting atmosphere as well as soft, delicate and textured fabrics.

② 主卧的效果图。
Rendering of the master bedroom.

月色星辰，静水流深

主卧呈纵深走向，以"进"为轴，设计强调了光线与空气的对流。夜色渐浓，若遇上清风揽月的时刻，卧室化身最佳的观景台。居住者在质朴素雅的氛围中凭栏望月，饱览城市夜色。沉淀心灵，回归静谧，这是为居住者献上的一处独有的空间体验。

Moonlight stars, tranquil waters

The main bedroom develops in depth, emphasizing the convection of light and air. When the night is getting dark, in the event of a breeze, the bedroom would be the best viewing platform, where residents can enjoy the moonlight and the city view in a simple and elegant atmosphere. A serene place for contemplation would be a rare moment for urban residents.

❹ 直接嵌入墙体的木门巧妙地隐藏了步入式衣帽间和卫生间，增加空间的序列感和整洁感。
The wooden door directly embedded in the wall tactfully hides the walk-in cloakroom and the bathroom, increasing a feeling of sequence and cleanliness of the space.

❺❻ 造型简洁的床头灯和细节处的陈设品充满现代精致感。
Modern refinement is filled in the simple bedside lamp and the details of furnishings.

❼ 卫生间的效果图。
Rendering of the bathroom is shown in the figure on the left.

❽ 干湿分离的卫生间确保安全和舒适。
Safety and comfort can be ensured through separating the wet and dry bathrooms.

关键词：自然、海景、休闲
Keywords: nature, sea view, leisure

建筑设计 / 筑博设计股份有限公司
景观设计 / 贝尔高林国际（香港）有限公司
建筑面积 / 330,000 平方米
委托范围 / 硬装及软装设计
委托面积 / 售楼部2,882 平方米
　　　　　别墅B户型 365 平方米
　　　　　别墅C户型 432 平方米

环海之滨同安，她的万千美丽全部都融汇在海里，以海的浪漫为世界所熟知。同安新区东连南安，西接长泰，西南与厦门郊区毗邻，东南隔海与金门岛相望，亦是著名的侨乡和台湾祖籍地。今保利叁仟栋坐落于此，坐拥7.9公里浪漫海岸线，东南俯瞰湾海，背依美人山，与海之厦门的结合就是最完美的邂逅。内揽新加坡贝尔高林滨海风情园林，外看优美海景。项目规划高层、别墅、商业等，倾力打造海岸墅区大宅，以超高的使用率和全海景感官，收纳稀缺海岸线。

Tong' An, a city surrounded by sea, is known to the world due to the romance of sea, whose beauties are gathered in sea. Tong'An New District is connected to Nan'an in the east and Changtai in the west, which is next to the suburb of Xiamen in the southwest and separated with the Jinmen Island by the sea. It's also a famous hometown of overseas Chinese and an ancestral home place for the Taiwanese. Now, Poly Sanqiandong is located here with a romantic coastline of 7.9 km, which can overlook the bay and the sea in the southeast; and it's backed by the Meiren mountain. It's a perfect meet with the combination of the sea in Xiamen. A garden with coastal sentiment of Belt Collins, Singapore can be seen internally, while the beautiful sea view can be seen externally. High-rises, villas, and business, etc. are planned in the project to spare no effort in building a large residential building in the coastal villa area, so as to embrace the scarce coastline with high utilization and full sea view.

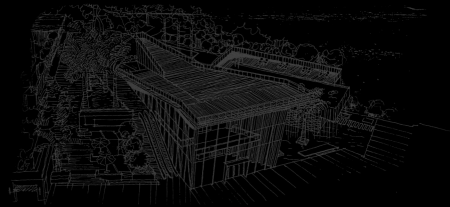

❶

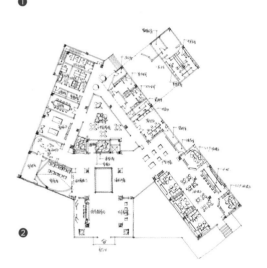

❷

坐落鹭岛的帆船

"面朝大海，春暖花开。"厦门同安叁仟栋项目，坐落于厦门深海旁，整体楼盘建筑如停靠于此的帆船，蓝色大海的静谧让途径此地的人留恋。因此在打造不同空间的室内场景时，结合建筑自身条件，设计师将不同的大海元素注入空间当中，用空间语言表达对大海的爱。用简单的空间装饰，打造舒适的居住空间，让人忘却工作的疲惫，尽情享受休闲时光。

❶ 一层平面图。
  Frist floor plan.
❷ 地下一层平面图。
  Minus one floor plan.
❸❹ 帆船似的建筑外观形态，像是载着侨胞们回乡的船。
  The sailboat-like building appearance is just like a boat carrying overseas Chinese back to their hometown.
❺ 售楼部沙盘区的吊灯，如海上翻涌的浪花。
  The ceiling lamp at the roof of the model is like surges on the sea.
❻❼ 奢华的别墅区设立了户外的阳台，居者可在此欣赏一线的海景。
  Outdoor balconies are set in the luxurious villa area; residents can have an outdoor balcony where residents can enjoy the front-line sea view.

Sailboats in Egret Island

Located by the seaside of Xiamen, Xiamen Coastal Mansion is just like sailboats at the harbor, enchanted by the serene ocean and unwilling to leave. Therefore, with such love for the ocean, the designer makes the most of the "ocean" element and properly integrates it into different settings. Simple furnishings make cozy living space, allowing people to leave fatigue behind and just enjoy life.

伫立的风帆

伫立于海旁的建筑，外形酷似迎风的风帆，独自在此感受来自海边的问候。转至室内，设计师将蓝色的静谧融入于售楼部的设计之中，在室内随处可见的蓝色元素，以沙盘区为中心点将功能平均的分布于售楼部的两侧。

整体空间以木质材料作为顶上的装饰，纵横分明；地面的大理石纹路与其相呼应，来访者顺应纹路走入室内。接待处设有项目的模型，可供宾客了解项目情况，接着则是莫大的沙盘区，将这个叁仟栋的规划缩影展示于人前。沙盘区的左侧是水吧和洽谈区，家具整体的排列软装则以白色为主，蓝色为辅。右侧则是VIP的洽谈室和娱乐区，在进行商务活动的同时亦可在此空间中休憩娱乐。去售楼化的设计，让空间的功能更加丰富。

❶

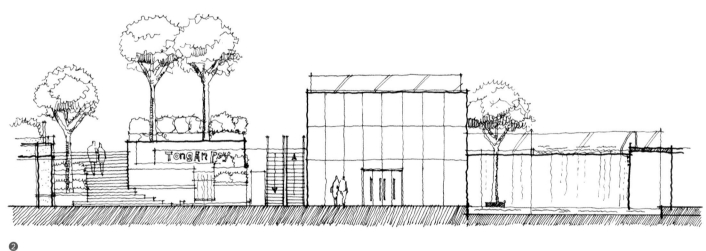

❷

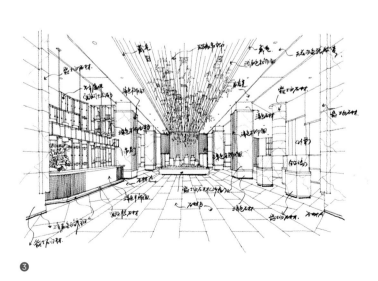

❸

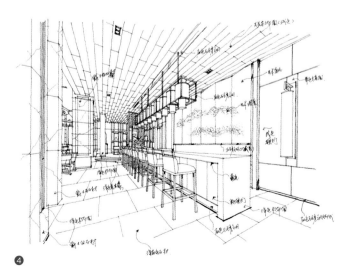

❹

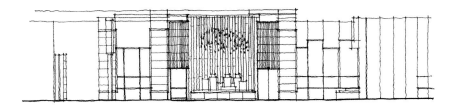
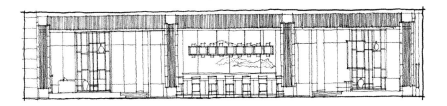
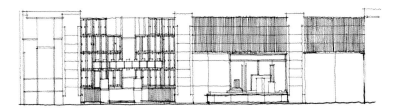

Standing sailboats

Buildings by the seaside are just like sailboats in the wind, embracing the ocean. Inside the room, ocean blue can be seen everywhere at the sales department where sand table is at the center and other function areas are evenly arranged on both sides.

The whole space uses wooden material for the top with clear-cut horizontal and vertical layouts which echo the patterns of the marble on the floor and visitors can follow the patterns into the room. The reception provides models of the project for reference. After the reception is a huge sand table which presents a vivid miniature of the whole project. On the left of the sand table is a bar and negotiation area whose furniture features in white with blue as supplement. On the right is VIP negotiation room and recreational area where people can relax and have fun while working. "De-marketing" design creates more functions for the space.

❶ 迎风的风帆。
Sails towards the windward
❷ 售楼部建筑景观剖面图。
Architectural section of the sales department.
❸ 在大堂的选材上，设计师巧妙的选用肌理较深的石材，在售楼部中展现水纹的肌理。
Stones with deep textures are adopted for the materials of the hall, while the water texture is displayed in the sales department.
❹ 水吧区设计构想。
Design concept of the water bar area.
❺ 售楼部各空间立面。
Spatial facades of the sales department.
❻ 售楼部沙盘区手绘图。
Sketch of sales department sand table.

① ④ 海浪的波纹化作空间的语言呈现于空间当中，犹如伫立的风帆。
Waves are turned into spatial languages to present in the space, just like a standing sail.

② 水吧区以错叠错的山峰图为背景。
The water bar area is set at the background of fluctuating mountain peaks.

③ 售楼部签约室装饰风格偏沉稳。
The decoration style of the signing room of the sales department is steady.

⑤ 通过立面图可以发现整体空间的线条感十足，充满了清爽干练的美感。
It can be seen from the vertical view that the whole space is linear full of refreshing and sophisticated beauty.

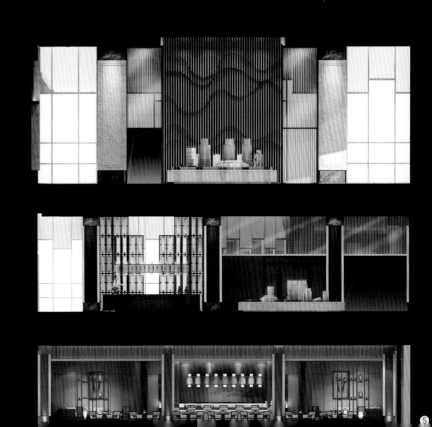

1. 金色的石纹地砖似沙滩上的流沙。
   Golden floor tiles with golden textures are like flow sand on the beach.
2. 3. 7. 综合考虑销售中心的后续功能，将多功能影院独立布局，以便空间复合使用。
   Considering the follow-up functions of the sales center, a multi-function theater is arranged independently for composite use of the space.
4. 5. 6. 设计师将大海的元素融入到空间当中。
   Sea elements are integrated into the space.
8. 蓝色与白色相间的VIP休息室。
   Blue-white VIP lounge.

复古爵士空间—B户型别墅

爵士的气质在经典、突破中游弋，经典的黑、白色，融入木材和金属、张扬的皮质等元素，在陈设艺术品的配合下，让两者完美碰撞。橘子橙色不经意地在黑色和白色之间流露，在色彩上达成一个鲜明的配置，空间顿生几许明艳。

复古，不代表守旧或固化。夜深时刻，啖一口雪茄，酌一斟美酒，品读生活。木质的天花暗藏暖光照明，与木质地板相映衬，给整个房间营造出静谧的气氛。窗外一缕缕阳光以肆意的角度照亮空间，包容的空间感，添置一张桌球台更体现主人的个性与爱好兴趣，无论是家人聊天还是朋友聚会将是一个理想的场所。

❶ B户型客厅手绘图。
Freehand sketching of the living room of the B-type apartment.
❷ 建筑外观图。
External view of the building.
❸ 橘子橙色在黑色和白色之中游走，色彩的交错令空间不再单调。
Orange is added in black and white for embellishing.

1

2 3

Retro jazz style—Villa B

Jazz temperament cruises in classics and breakthroughs. With the installation of works of art, the classic black and white, blending in elements of wood, metal, leather, and so on, work perfectly with one and another. And the orange naturally reveals between black and white, adding a more vibrant flavor to the room.

Retro, does not mean conservative or solidified. A sip of cigar, a glass of wine, and some peace of life make a perfect night. The wooden ceiling is lit up with warm lighting, and the wooden floor is lined up to make a tranquil night. During daytime, rays of sunshine illuminate the space through windows. A pool table in a vast space would highlight the owner's personality and hobbies. Be it a family chat or social gatherings, it will make a perfect place.

❶❷ 开扬的设计，窗外的景色一颗在此一览无遗。
Scenery outside of the window is clear with the open design.
❸ 中空的地下层变身家庭成员的娱乐场，也不乏与上层形成互动。
The hollow subterranean layer can be transformed into a recreation place for family members with interaction remained with the ground floor.
❹❺❻❼ 异形的家具为空间增色不少。
Furniture in special shape really give highlights to the space.

❶❷❸ 晨曦的一缕阳光洒满主卧，点亮新的一天。
Sunshine in the early morning can fill the master room to light up a new day.

❹❺❻ 主卧的卫生间朝向大海，设计师将窗外的美景引入室内。
The bathroom of the master bedroom faces the sea, as the designer would like to introduce the fine view out of the window into the room.

❶

蓝色的灵动空间——C户型别墅

水有千态，动静皆宜。设计师将设计风格方面界定为简洁大方的现代新简约主义风格，体现主人的个性与品位。本案环海户型，让心灵与大海更接近，观景阳台与水亲近，能涤荡忙碌奔波后的余尘，令心情面貌焕然一新。空间基调以黑、白色为主，搭配一些绿色植物衬托，使空间充满惬意、轻松的气氛。装饰材料以亮面木材、简约线条，搭配造型简洁的灯具，带来既前卫且不受拘束的感觉。

设计师充分考虑到室内与室外空间的交流，因势利导的设想，让明媚的阳光透过落地窗户照射入内，简洁的家具搭配湖水蓝的点缀，显得空间宽敞明亮。主人房同样以黑、白为主色调，引用线条和不规则图形，搭配木制墙面和家具，朴实无华，隐藏在天花的暖光灯为房间营造出温馨舒适的感觉。

地下室空间延用大量的木材装饰，在黑与白之间隐约透露了主人的品味，注重生活品味的同时还注重健康时尚；保利叁仟栋用简洁的心与海亲近，感受那片广阔与舒适的生活体验。

❶ 客厅创作手绘图。
Freehand sketching of the living room.
❷❸ 空间各处的绿植点缀。
Green plants decorated around the space.
❹ 阳光洒落的温暖客厅。
The living room is warm and filled sunshine.

① ②

③

④

⑤ ⑥

❶❷ 客厅中全玻璃幕墙的设计，光线在室内空间中自由流转。
A full glass curtain wall is designed in the living room, so that the light is free to flow in the indoor space.

❸ 简约温馨的餐厅。
Simple and warm restaurant.

❹❺❻ 常年在外生活的侨胞，有下午茶的习惯，用餐则使用西式的餐具。
Overseas Chinese, who live abroad all year round, have the habit of afternoon tea, and use western-style tableware for meals.

❼ 客厅和楼梯不再是独立的两个整体，墙身嵌入玻璃让两个空间有了互动。
Instead of being independent, the living room and the stairs are interacted through embedding a glass into the wall.

❽ C户型别墅的建筑剖面图。
Architectural section of the C-type villa.

Blue space—Villa C

The designer adopts modern new minimalist style (simple and elegant), aligned with the owner's taste. This seaside portfolio-- balcony near the sea , brings the heart to the sea and refreshes one's mind. With black and white as the main tone, the space introduces some green plants to add some liveliness and coziness. The application of glossy wood, simple lines, and neat lighting makes the room free and fashionable.

In consideration of the indoor and outdoor interaction, the designer cleverly lets the sunlight shine through the French window. And along with the simple furniture embellished by lake blue, the room appears more spacious and bright. The main bedroom is simple and nice with black and white as the main tone, coupled with lines and irregular figures, wooden wall and furniture. The warm lights hidden at the ceiling help to create a cozy feeling.

The basement continues to use plenty of wooden furnishings. The black and white reveals the owner's pursuit for quality life, health and fashion.

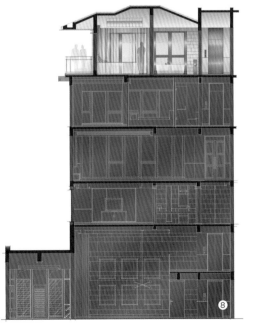

① 首层户外花园。
Outdoor garden on the ground floor.
② 顶层天台户外风光。
Outdoor scenery at the rooftop.
③ 设计师注重室内外的互动，晨起落日，居者都可在此观赏美景，细数年华。
Interactions between indoor and outdoor are emphasized by the designer. Residents can enjoy the sunrise and the sunset here.
④⑤ 主卧卫生间设计通透明亮。
The master bathroom design is bright.
⑥⑦ 现代简约的主卧空间，软装以蓝色作为主色，与户外的蓝色的大海相呼应。
The modern and simple master bedroom is decorated primarily in blue, echoing to the blue sea outdoor.

❶❸ 对于设计师而言，地下空间是另一处生活的舞台，不再仅限于地库的功能，可在此接待客人，亦可独享自身的家庭时光。
As for the designer, the subterranean space is another living stage rather than limiting to basement function, where can be also used for welcoming guests or enjoying family time.
❷ 私属的功能区更多布局在相对独立的区域。
Exclusive function areas are arranged in independent areas.
❹❺ 开放式的地下层，保证了空间之中光与气息的流动。
The open subterranean layer ensures the flow of light and air in the space.

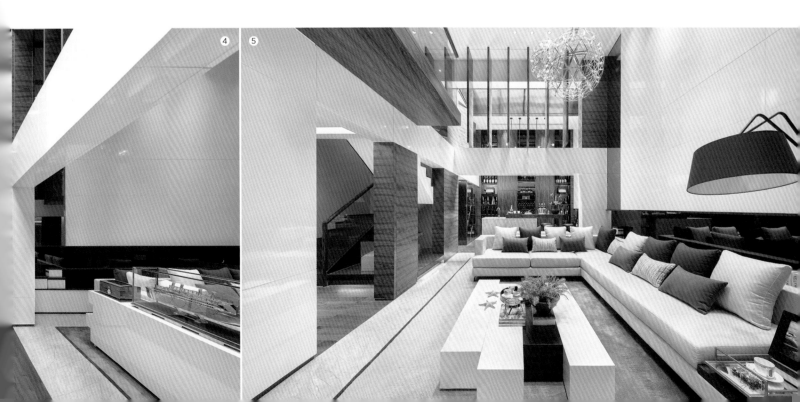

地标建筑
LANDMARK BUILDING

标志性建筑是一个城市的名片和象征。室内设计不但要准确把握城市气质营造氛围，也要以整体性思维合理巧妙规划大体量建筑内部的功能布局，使其实现建筑内外美感的统一，以及功能与美学的统一。

The landmark building is a city's name card and symbol. For its interior design, the designer should not only accurately grasp the urban temperament to create a proper atmosphere, but also rationally and cleverly plan the interior functional layouts of the building as a whole, so that the unity of the interior and exterior aesthetics and the unity of functions and aesthetics can be realized.

三亚保利财富中心
广州保利国际广场
广州保利天幕广场
广州华南国际港航服务中心
珠海中冶盛世国际广场
Poly Wealth Center, Sanya
Poly International Plaza, Guangzhou
Poly Skyline Plaza, Guangzhou

三亚保利财富中心
POLY WEALTH CENTER, SANYA

关键词：阳光海岸、自然、全景公寓
Keywords: sunshine coast, nature, full-view apartments

建筑设计 / 美国GP建筑设计有限公司
景观设计 / P Landscape Co., Ltd
建筑面积 / 210,000 平方米
委托范围 / 硬装及软装设计
委托面积 / 游客中心 765 平方米
　　　　　顶层复式公寓 1685平方米
　　　　　酒店式公寓 62441.7 平方米
　　　　　酒店会所 545.5平方米

海洋令人着迷。白帆饮饱了风，船舫满载抱负，于湾峡津渡纵力出发，径直驶向未来。勇于挑战、追求自由的冒险精神自浪潮中诞生，随风雨传递，将热血嵌入后世开拓的图谱。它是大航海时代留给后世最宝贵的遗产。

眼前的阳光雨露、沙滩白浪，对于在事业和生活中劈波斩浪的勇者而言，不仅仅是放松身心、度假玩乐的好去处，更是一片财富增长的热土。

国家海岸，城市远景，这是诚挚礼献给冒险家和人生赢家的一处眺望未来的灯塔。

Ocean is always enchanting. With an adventurous spirit, sailors embrace challenges and set sail, weathering through rains, winds, and waves... and finally opening up a new world. And this is the most precious legacy the Age of Discovery has left for later generations.

Sunshine, beach, waves... these for elites are not only a place to relax and go on vacation, but also a spot for growing wealth.

With the national seashore and promising city prospects for adventurers, this property is like a beacon, lighting the way forward.

❶ 图示为整体项目效果图。项目坐落于海棠湾畔,与蜈支洲岛隔海相望,占据国家黄金海岸线,尽享顶级的海景资源。
Rendering of the overall project is shown in the figure. The project is located at the shore of Haitang Bay and separated with the Wuzhizhou Island by the sea, which enjoys the top-level seascape resources occupying the golden coastal line of the country.

❷ 裙楼13层天面规划了达130米的空中无边际游泳池,将全国绝无仅有的热带优质资源发挥得淋漓尽致。
A 130-meter overhead boundless swimming pool is planned on the 13th floor of the podium building to give full play to the unique tropical resources in the country.

❸ 三亚·财经国际论坛永久会址,以"磐石"为理念,象征年度集聚的智库资源与经营企业家群。
With a concept of "huge rock", the permanent site of holding the Sanya· Finance and Economics International Forum symbolizes the annual gathering of think-tank resources and entrepreneurs.

保利财富中心位于三亚海棠湾海岸大道东侧，为国家黄金海岸重点项目，囊括三亚·财经国际论坛永久会址、61栋全海景奢华别墅、超五星级瑰丽酒店、216米地标性海景公寓四大业态，总建面约21万平方米。其中塔楼总高度210米，于2018年4月底封顶，是海棠湾目前最高的建筑。它月牙形的建筑造型以及顶部玻璃灯塔的设计模拟出一个巨大的灯塔。三亚保利财富中心集高端会议及多功能中心、高端商务酒店为一体，不但是三亚乃至是整个海南岛的标志性建筑，而且将作为三亚财经国际论坛的永久会址，照耀一湾海景。

Located at the east of the coastal avenue of Sanya Haitang Bay, Poly Wealth Center is a key project in national gold coasts portfolio with an overall floorage of about 210,000 square meters. Its tower building, completed at the end of April 2018 and with a height of 210 meters, is the highest building in Haitang Bay. Its crescent-like outlook along with the glass beacon-like design at the top makes it a huge beacon. Serving as a high-end conference and multi-functional center as well as a high-end business hotel, Poly Wealth Center is not only a landmark at Hainan Island, but also a permanent site of Sanya Forum.

❹ 落成后的财富中心即将成为区域鲜明的地标建筑。
The completed wealth center will become a distinctive landmark building in the region.
❺❻ 项目的建设过程。
The construction process of the project.

❶❷❸ 游客接待中心采用浪漫的新亚洲设计风格，吊灯和地板的设计分别对应着金色的海浪及岸边沙滩。
A new romantic Asian design style is adopted in designing the visitor center. The designs of the ceiling lamp and the floor correspond to the golden waves and the sand near the shore.

❹ 温暖的余晖为建筑群镀上一层耀目的金黄。
The warm sunset glow coats dazzling golden on the buildings.

❺ 别墅群的建筑及室内装饰均采用出简约利落的风格，无敌的海景是最为显著的价值体现。
The architectures and interior decorations of the villa groups are designed simply and neatly style; while the invincible seascape is it's most remarkable value.

扑面而来的东南亚风情，无处不在地诉说着海洋的故事。接待中心位于"磐石"附近，室内设计上也与之相应，提出凝练的直线条，使空间看上去沉稳干练。东南亚风情的家具及装饰品、温馨淡雅的中性色彩，是我们对自然、原始风格的揣摩和再读，务求让崇尚自然、喜爱热带文化的成功人士享受到宾至如归的感觉。

Located close to the "monolith", the reception center echoes it in design, applying straight lines and making the room composed and neat. Carefully studying natural and primitive styles, we introduce southeast Asian-style furnishings and warm neutral colors, allowing elites that favor tropical culture feel like home.

复式C户型-首层
Double C - top floor

复式A户型-首层
Double A - top floor

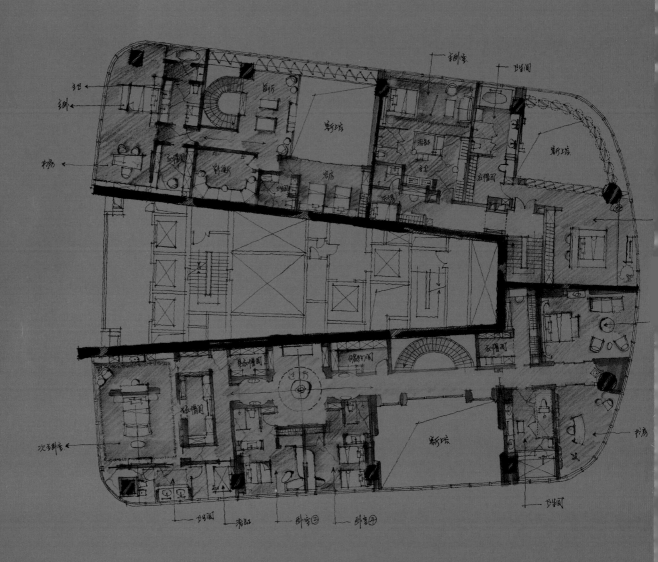

我们利用建筑的特点以及顶层环景露台的鸟瞰优势,我们划分了三个复式户型,并针对潜在的客群特性分别为其定制主题。每个户型空间也因此拥有了自身独一无二的个性与形象。
Considering the features of the building as well as the advantage of a panorama terrace with the bird's-eye view, we introduce three types of duplex houses and tailor-make themes for specific client groups. Each house type boasts its unique trait and image.

三种不同户型的顶层复式公寓给予顶级冒险玩家全新的感官世界。每种空间都意味着一种生活态度。对于顶级玩家来说，度假既是玩乐，也是社交活动。他们对环境敏感且追求极致。居所里的所有材质必须上乘且品位卓越，展现独特的个性气韵。登高望远：独揽270°一线海景时将会收获与众不同的视野——此时的海平线不再平直，而是展露出地球边际微妙的曲线与弧线。如此一来，无论置身居所的何处，都能领略临海观澜的独特滋味。

Duplex penthouses with three different house types offer a brand new world for top adventurers. Each house type represents a life attitude.Top players going on vacation are for fun as well as for social. Their requirement for space is sensitive and perfection-oriented: all texture in the house should be superior and with taste, coincided with the owners' personalities. The unique 270° sea view offers a unique view where the sea level is no longer straight but with subtle curves. In this way, wherever you are at the house, you can have an immersive sea-watching experience.

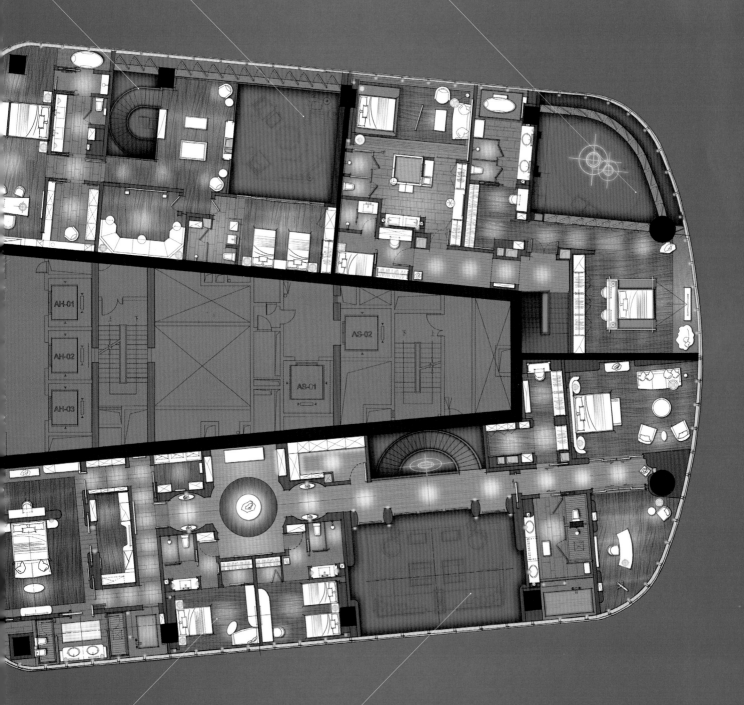

中空　Building hollow

复式B户型-上层　Double B - top

中空　Building hollow

复式B户型-首层
Double B- top floor

复式C户型-上层
Double C - top

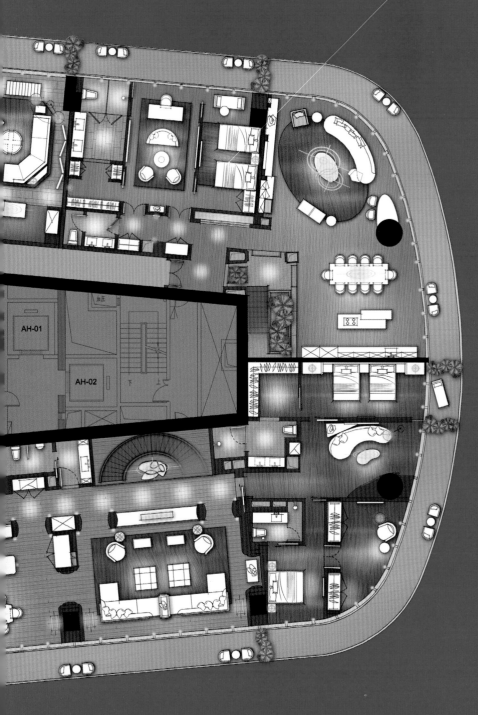

复式A户型-上层
Double A - top

与奢华游艇有关的一切

游艇俱乐部兴盛于18世纪的欧洲,逐渐演变成社会上流人物的聚集地。现代社会的俱乐部文化,已经从原有的简单功能发展到集餐饮、娱乐、住宿、商务等多功能于一体。以奢华游艇为原型,SNP在超高层再造了一个游艇俱乐部。

A户型呈深进形态,从直达电梯的出口处客人们即可"上艇"出发。进入内部,纵贯整个空间的通高玻璃幕墙最为惊艳,海洋的千姿百态在界限模糊的状态下展露无遗。双层中空设计兼具社交和生活功能,布置轻松且时尚。白日时分可以悠闲观海,入夜后这里化为热闹的派对舞池,电子火炉和垂吊到地面的巨大吊灯共同营造出梦幻的气氛。放一首爵士,踏着海浪和繁星闪烁的节奏翩翩起舞,尽情享乐到深夜。

Everything related to luxury yachts

Rising in the 18th century in Europe, yacht clubs gradually turn into a gathering place for the upper class. Modern club culture has changed from simple functions to multi-functions including catering, entertainment, accommodation, business, and etc. With luxury yachts as a prototype, SNP builds another "yacht club" at the super high-rise buildings.

House type A is in an in-depth shape. Guests can "get on the yacht" by elevator at the entrance. Inside the "yacht", the floor-to-ceiling glass curtain wall is the most stunning part where different sides of the ocean can be seen. It is a double-deck room with multiple functions for social and daily life in a free and fashionable style. People can enjoy the sea views at daytime; while at night it turns into a dancing floor with the electronic stove and the to-the-ground chandelier helping to create a dreamy atmosphere. People dance to the jazz among the "waves and shining stars" till midnight...

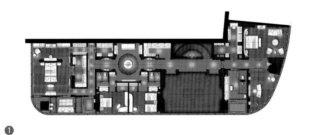

❶

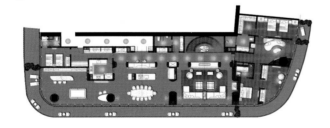

❷

❶❷ 顶级私享A户型平面。
Plan of the top-level private A-type apartment.
❸ 超高层建筑中双层中空大客厅既保证采光,同时也将景观作为最好的装饰。
The double-layer hollow living room in the super high-rise building is well decorated the landscape on the basis of ensuring day lighting.

① 入户玄关干净的立面显得两尊雕塑更加趣味横生。
The two sculptures look more interesting with comparison to the clean facade of the hall way.

② 二层过道的艺术区，壮阔人生的星辰大海都在此处相遇。
Stars and sea for significant life are met in the art district on the aisle of the second floor.

③ 餐厅与小酒吧及开放式厨房相连，意味着更加丰富多变且体验一流的就餐场景。
The dining room is connected to the small bar and the open kitchen, presenting a rich and varied dining scene with the first-class experience.

④ 延续海洋的设计概念，素雅的色调加入了柔和的天蓝作点缀；套房内部各种品质细节重现漫步阳光海滩的度假情调。
The sea design concept is extended. Soft sky blue is added in the elegant color for embellishing; the holiday atmosphere of walking on the sunny beach reappears with all kinds of quality details in the suite.

玄关墙的背面被设计为西厨的操作台，引进的意大利知名橱柜品牌让烹饪过程更加得心应手。沿着旋转楼梯拾级而上，过厅如同一个小小的艺术展示区。天幕上铜制的漫天星宿图，对应着地面麦哲伦的航海地图；水晶装饰品恍惚间投射着一个久远的故事。主人套房开敞面海，宽大的落地窗视野开阔，面朝大海，盥洗室拥有浴室、按摩浴缸以及齐全的桑拿设备。卧室内波浪纹的地毯、浴室湿区的鹅卵石和木地板，令尊贵的主人足不出"船"，继续享受海边浪漫写意生活。

The back of the hallway wall is turned into an operation floor equipped with cabinets from the famous Italian kitchen brand Veneta, making the cooking more enjoyable. Ascending along the spiral stairway, one can see a small art-zone-like corridor. The ceiling is coverd by a copper-made starry map, echoing the nautical map by Magellan on the ground. The crystal furnishings reflect the light and seem to tell an age-old story. The master suite faces the sea with large French window, equipped with sea-facing bathroom of multi-functions like spa and sauna. The carpet with wave patterns in the bedroom, cobble stones in the bathing area of the bathroom, wooden floor... all of these allow our noble guests to enjoy a cozy life by the sea without stepping out of the "yacht".

天风海涛，从此静心

洋溢着现代主义风格的B户型用白砂岩及白色的混凝土唤起对海洋的感知。入户玄关不再直白地展现奢侈的海景，而是利用立柱构造景框，将餐厅和远处的海天一色容纳其中，耐人寻味。穿过玄关，即来到了相连的餐厅和客厅。客厅独占一片洁白幕墙，吊灯造型简约独特，幕墙底部木饰面和木地板充满了自由悠闲的度假气息。

Winds and waves calm the mind

The house type B of modernism uses white sandstones and white concretes to arouse people's consciousness of sea. Instead of displaying the sew view in a straightforward and plain way, the hallway uses standing columns to put the dining hall and the distant sea view within a thought-provoking frame. Through the hallway comes the unobstructed dining room and living room. The latter hangs a special chandelier and occupies a whole white curtain wall whose bottom applies wooden veneer echoing the wooden floor, delivering a care-free vacation style.

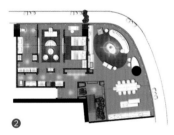

❶❷ B户型首层、二层平面图。
First floor and second-floor plans of the B-type apartment.

❸ 入口玄关加入了3米的水族箱景观是与户外的海景别具巧思的呼应。
A 3-meter aquarium landscape is added in the hall way to echo with the outdoor seascape.

❹ 独具匠心的灯饰搭配餐桌、沙发，令餐厅客厅形成互动。
The unique lighting with a dining table, sofa generate an interaction between the living room and the dinning room.

游走在公寓各处，设计师引导自然光线跃动穿梭，因而整体空间通透明亮。舍弃繁杂，设计师以轻盈、简约、当代的手法，布置空间的各个功能细节，呈现世界国际的空间设计潮流之余，又为主人带来明净的视觉体验。度假，最重要的还是彻底放松下来，享受大海带来的安静和纯粹。

The designer makes the most of the natural light, making every part of the apartment clear and bright. With light, simple and modern techniques, every detail of the space demonstrates up-to-date fashions of international interior design and brings neat visual experience to people. For vacation, what matters most is to fully relax and enjoy the peace of sea.

❶ 连片的海色是尊贵主卧最好的装饰艺术。
Consecutive sea color is the best decorative art of the distinguished master bedroom.
❷ 二层客厅概念草图。
Concept sketch of the living room on the second floor
❸ 善用二层客厅空间，安排书桌与围拢式沙发组，令工作交谈更加舒适。
The space in the living room on the second floor should be used properly through arranging a desk and a sofa group, so that work and conversation can be performed in a comfortable way.

❹ 卧室床具融合了书桌的功能，通高的柜体又特别纳入了酒吧的功能，微醺慵懒的生活细节。
Beds in the bedrooms have the desk function; and the full-height cabinets consist of the bar functions, giving you living details of slightly drunk and relaxing.

❺❻ 浴室采用来自国外知名品牌Dorn bracht及汉格雅思的卫浴配件，入墙式化妆镜、全自动感应水龙头等均外观内涵俱佳。
Bathroom accessories from well-known foreign brands Dorn Bracht and Hansgrohe are adopted in the bathrooms. And the wall-in vanity mirror and automatic sensor faucets are perfect in both appearances and connotations.

❼ 卧室内部手绘图。
The freehand sketching of the interior bedroom.

海底百米，水下的神秘世界

沉船的传奇如同一个引人入胜的宝藏，吸引那些低调却富有活力的爱好者只身深潜百米，探索沧桑神秘的人类文明遗迹，接触那更有质感的过去。这个过程体现着人类与自然之间的博弈和共融。我们用国际工业风低沉的色彩和空间格调，向这些全身心投入所爱的勇士和冒险家致敬。

宽长的玄关尽头，旋转楼梯静静伫立。它由特别处理的钢板构筑而成，还原了沉船上的斑斑锈迹。楼梯下的一个浅浅的白沙池，好像被雕塑点晕开一圈圈的涟漪。这种视觉表现手法令人浮想联翩，让人不经意间代入到情景中去。

100 meters below the sea—a mysterious world

A sinking ship is like an appealing treasure, attracting those low-key but enthusiastic fans to dive on their own to discover the ancient and mystic relics of human civilization. This process shows the battling and compromising between human and nature. We use dark tones of industrial style to pay our tribute to these brave men.

At the end of the hallway stands a spiral stairway made of special steel plates which restore the rusty stains of a sinking ship. Down the stairway is a small patch of white sand, just like circles of ripples touched by a sculpture. People might unconsciously immerse themselves into the scene on account of this thought-provoking design.

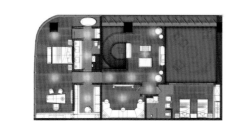

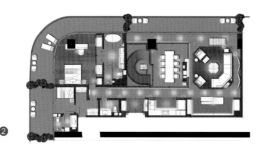

❶❷ C户型二层、首层平面图。
Plans of C-type apartments on the first floor and the second floor.
❸ 大面积的石材以及铁丝网等视觉元素，从入户玄关开始便展示出国际工业风的特点。
Adding visual elements such as wide-spread stones and wire meshes can show the features of international industrial style from the hall way at the entrance.

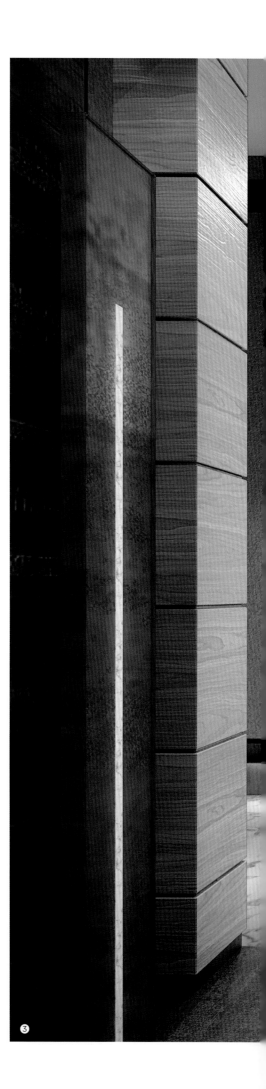

❶❹❺❻ 用以烘托气氛的电子壁炉、灵感源自海上探照灯及旧式潜水设备的落地灯饰、艺术品，更是诉说着空间每一个充满惊喜的细节。
An electronic fireplace for foiling the atmosphere and artworks inspired by sea searchlights and old diving devices.

❷❸ 与楼梯间相连的餐厅同样可以看到浩淼的大海。
The dinning room connected to the stairs can also see the vast sea.

毫不炫技的Loft设计风格不但令空间资源得到合理的分配，还让上下两层空间加强了视觉的互动和交流。软装陈设展示出典型的美式风格。纹理自由灵动的大理石和质感厚重的家具，既营造了海底下安静且微妙的景象，也勾勒出主人踏实稳重的生活态度。

The Loft design style contributes to not only reasonably allocating space resources, but also enhancing the visual interaction and communication between spaces in the upper and lower layers. A typical American style is presented with display in the soft decoration. Marbles with free textures and furniture with heavy texture can not only build a quiet and subtle scene under the sea, but also outline the owner's steady living attitude.

❶❷❸❹ 二楼主卧套房房布局简洁，功能实用；隔间之间恰到好处的玻璃设计增添虚实相生之感。
The master bedroom suite on the second floor is functional in simple layout; a sense of false or true complement is added with the glass design right between the compartments.

❺❻ 一楼套房巧妙地化用原建筑的立柱作为电视墙和书桌。
Pillars of the original building are tactfully adopted as television background and desk on the suites of the first floor.

❼ 图为主卧的衣帽间。
The cloakroom of the master bedroom is shown in the figure.

无处不在的的灰棕色调和硬朗的空间线条，散发着阳刚的气息。映衬着阳光与海滩，卧室也仍是透露着如礁石般安稳、坚固的色调。大量的原木、石材等元素强调空间自然质朴的底色。当光影流转，悄然入梦，这些玩家会否想起从海底仰望时看见的粼粼波光呢？

The taupe brown tone and tough lines show an air of masculine. Facing the sunshine and beach, the bedroom still applies the color tone which implies steadiness and firmness.

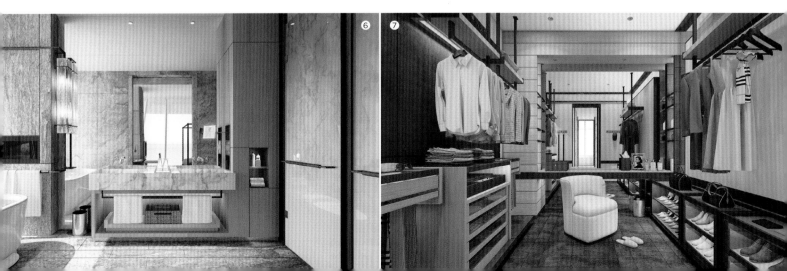

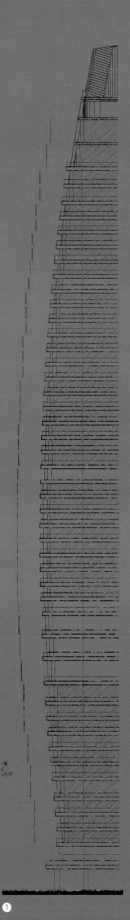
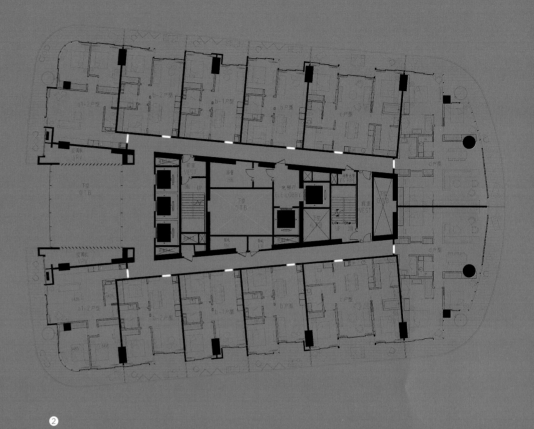

① 建筑的玻璃幕墙立面以每三层为单位作退缩处理。
The glass curtain wall of the building is retracted by every three layers.
② 标准层平面图。
Plan of standard floors.
③ 经设计研究统筹后的户型分布一览表。
It's a list of apartment distribution after design research and coordination.

酒店式服务公寓围绕核心筒环型布置，采用类住宅平层设计，注重户型功能组织。借助BIM技术，SNP将标准层公寓划分为： A类9种， B类3种，c类5种， d类1种合共18种户型，横跨2-11、15-22、24-35、37-43楼层。我们成功解决了原建筑户型过多过杂（原户型类别多达50余种）的问题，帮助降低工程成本及后期的户型模型管理。在保证阳光气流畅通无阻的同时，户外借景的手法又完美保留一线无敌海景。

Service apartments are arranged around a core tube in a ring shape, applying residential-like designs and focusing on functional division of different house types. With the aid of BIM, SNP divides the standard floor into 18 house types (9 Type A, 3 Type B, 5 Type C, and 1 Type D) covering F2-F11, F15-F22, F24-F35, and F37-F43. We have successfully solved the problem of too many house types (more than 50) in the original building and helped to reduce the engineering cost and manage the house type model in the later period. The technique of borrowing scenery from the outside ensures the convection of air and light, meanwhile keeps the splendid sea view.

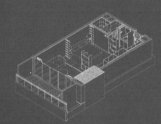

A户型（A family）
53.3m²

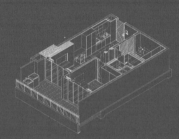

B户型（B family）
70.25m²

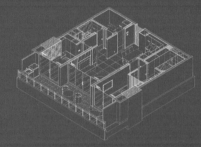

C户型（C family）
106.44m²

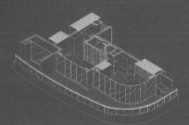

A1户型（A1 family）
80.3m²

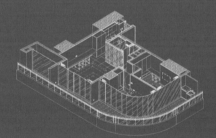

A2户型（A2 family）
89.04m²

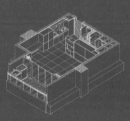

A3户型（A3 family）
57.98m²

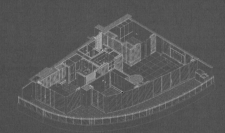

D户型（D family）
142.88m²

A5户型（A5 family）
45m²

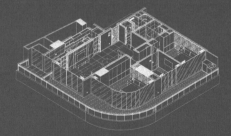

A4户型（A4 family）
145.45m²

明快简约，轻假期的安眠之处

户外室内装饰遵循浓淡相宜，自然朴素的原则，务求带来休闲放松、自由放飞的感受。置于客厅与卧室之间的纱幔窗帘飘荡，既轻柔地隔开两处空间，还能让人具体地感受到海风的轨迹，体验空间动静结合的美态。去风格化的设计手法，就如大海的藏露不显、收放自如，为主人带来纯粹、实用的空间体验和空间价值。

Simple and neat, the best place for care-free vacation

Exterior and interior furnishings stick to the principle of being simple and fitting, aiming to bring relaxing and care-free experience. "De-stylized" designs, composed and sophisticated like the sea, bring to the master authentic experience and value of space.

❶ 图为公寓的客厅。
The living room of the apartment is shown in the figure.
❷ 图为公寓的卧室。
The bedrooms of the apartment are shown in the figure.
❸ 充满白色优雅的陈设品。
It's full of white and elegant furnishings.
❹ 白色的渲染，蓝色调的点缀，度假公寓也可以充满居家感。
A vacation apartment can be filled with a sense of home through rendering in white and embellish in blue tone.
❺ 于阳台远眺远处的蜈支洲岛。
Wuzhizhou Island can be overlooked on the balcony.

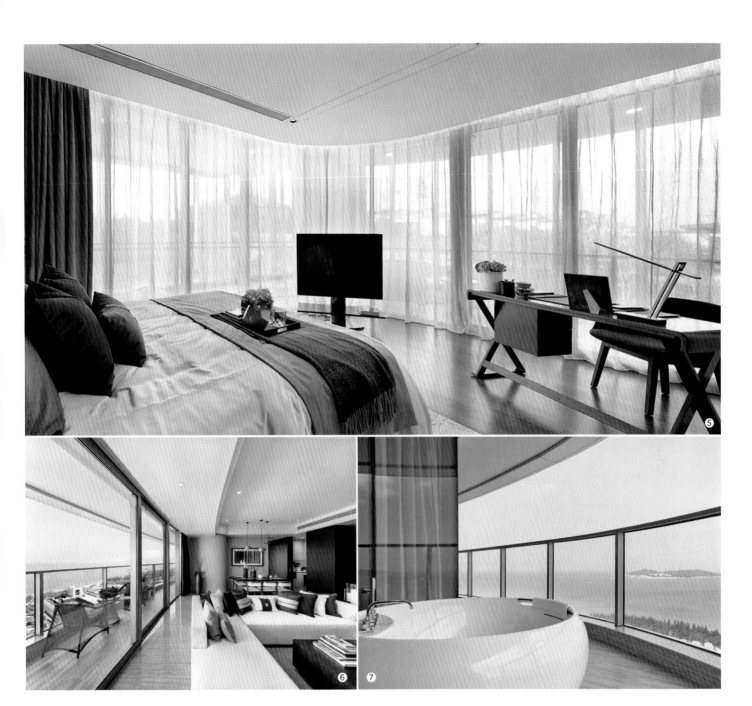

❶❷❸❹ 推拉门能够起到空间功能划分和隔音的效果，使用方便灵活，可减少对空间的浪费。
The sliding door can help divide spatial functions and sound insulation, which is convenient and flexible, reducing spatial waste.

❺❻❼ 环形阳台保证了景观资源的最大化。
Landscape resources can be maximized with the circular balcony.

公寓阳台采用了局部退缩的设计手法，局部阳台进深达2.4米，使建筑立面效果更显柔美的同时，让阳台的功能更为多样：例如打破原有的定式，将浴缸安排在阳台。沐浴于蓝空夜星之下，跟开拓人生蓝图一样，不再是遥不可及的梦想。

Adopting the technique of "partial retreat", part of the terrace stretches inside for 2.4 meters, adding mildness to the exterior façade and meanwhile diversifying the terrace's functions. For example, break the convention and bring the bathtub to the terrace. Bathing under the starry sky, like building one's own empire, is no longer an impossible dream.

热带独特的浪漫情怀

滨海地带的热闹，当然是要有声有色。木作既作饰品，也以重复的手法成为隔断，为会所大堂带来浓郁自然的热带气息。走廊立面保留凹凸错落的原木肌理，形象贴切且趣味十足。用餐区采用清新唯美的粉蓝色、白色作点缀，寓意着壮美的蓝天白云。自然色系在室内的运用，达成了户内户外色彩的统一，这也是设计师的巧思所在。

Tropical and romantic clubs

Carpentry is used as a decoration and a separation by means of repeating, bringing a tropical atmosphere to the lobby. The original concave-convex wood texture is reserved for the corridor facade, featuring an image appropriate and interesting. The dining area is decorated with fresh and beautiful pink blue and white, symbolizing the blue sky and white clouds. The use of natural color in the interior contributed to the unity of indoor and outdoor colors, which is also the designer's ingenuity.

❶❷ 大堂走廊的木结构立面充满趣味细节。
Interesting details are filled in the wooden structure facade of the lobby corridor.
❸❹ 日间的会所空间明亮且优雅。
The clubhouse is bright and elegant at the day time.
❺ 立面概念图。
Facade concept elevation.
❻ 会所平面方案一、方案二。
Floor plan 1 and plan 2 of the clubhouse.
❼ 用餐区充满了白色浪漫。
The dining area is filled with white romance.

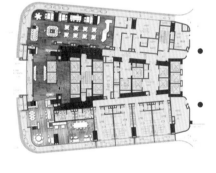
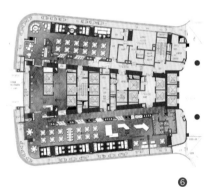

主体灯饰造型如同海上的鱼笼；辅助灯饰垂坠于空间各处，仿似在描摹深海中浮动的串串气泡或是海面随风起舞的微澜。这些形象应景又温暖明亮，烘托出浪漫优雅的氛围，让人不经意沉醉。繁星低垂的时分，海潮的呢喃与美食佳肴相得益彰。

The main lighting is like a fish pot at sea; its accessories scattered at the space are like floating bubbles or dancing ripples. Such warm and bright images, coincided with the occasion, delivers an air of romance and elegance.

❶ 昼夜交替，会所的视觉形态也随之发生变化。
The visual form of the clubhouse changes accordingly with the change of day and night.

❷❸❹ 延续海浪的形象，酒廊区的家具呈现出半围拢的形态，灯饰也以曲线、串珠等元素展现。
Furniture in the lounge area is presented in a semi-enclosed form to extend the wave image; meanwhile, lights are also presented in elements such as curves and beads.

❺ 渔民有用竹笼捕鱼的传统，设计师将这一意象化为了设计元素。
As the fishermen have a tradition of fishing with bamboo cages, the designer made it into a design element.

❻❼ 图为会所的户外就餐区，在此可观赏到海边迷人的夜色。
The outdoor dining area of the clubhouse is shown in the figure, where you can enjoy the charming night of the beach.

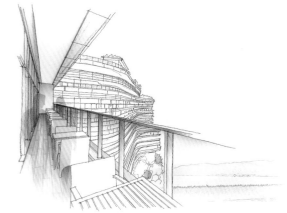

广州保利国际广场
POLY INTERNATIONAL PLAZA, GUANGZHOU

关键词：一线江景、创意空间、灵感之所
Keywords: full view of the river, creative space, inspiration

建筑设计 / 美国SOM建筑事务所
景观设计 / 美国SWA公司
建筑面积 / 19,600 平方米
委托范围 / 硬装及软装设计
委托面积 / 1,300 平方米

保利国际广场是保利发展倾力打造的重点项目，该项目位于广州琶洲会展区域，面临珠江，建筑面积195,000平方米。由南北两栋165米高的"超长板式"办公塔楼和东西两栋4层商业群楼围合而成，北望珠江一线江景，建筑和园林设计分别由美国SOM事务所和SWA景观公司担纲，是会展中心区域配套完善的国际五星商务写字楼。

设计师用当代的方法表达了对中国传统式园林的尊重，给人梦幻般的感觉，却又有实实在在的感觉。"SNP（尚诺柏纳空间策划联合事务所）自保利琶洲综合体项目之处，全程参与，在本案保利国际广场中，为业主实现了多项创作。

Poly International Plaza is a key project of Poly Real Estate Group, located in Pazhou, Guangzhou, near the Pazhou Convention and Exhibition Center, with a building area of 195,000 square meters. Facing the Pearl River, the plaza is an international five-star business office building with complete supporting facilities. It consists of two 165-meter-high "super long-board" office towers in the south and north and two four-story commercial buildings in the east and west. To the north, one can enjoy the full view of the River. The buildings and gardens are respectively designed by the US companies of SOM and SWA.

The designer expresses his respect for traditional Chinese gardens in a contemporary way, giving people a dreamlike feeling, but with real meanings." SNP (Sunny Nuehaus Partnership) has participated in the whole project starting from the Poly Complex Projects in Pazhou. In this case, it has realized many creations for the owners.

1

① 产品体验中心入口处。
The entrance of the experience center.
② 弧形的前厅,完整展示品牌沉淀的历史。
Arc-shaped front hall, a complete display of the brand precipitation history.
③ 橱窗内展现全生命住宅内的简易格局。
Inside the shop window, it shows the simple pattern inside whole life residence.
④ 产品体验中心接待处。
Reception of product experience center.

和悦展厅

秉承"美好生活同行者"的理念,保利发展一直致力研发和打造契合城市再生理念的人居生活空间,根据对新一代居住模式的分析,建立了产品体验中心,诠释了保利发展"和悦系"全生命周期住宅。

和悦展厅的设计以参观者的动线为基础,依次就产品历史、产品标准、社区建设等方面做分区展示呈现"和悦系"住宅产品。入口接待处一旁设置简易的橱窗,展现全生命住宅内的格局布置,让参观者对空间有初步的构想。走入前厅,弧形的展示墙上展示着产品的发展历史与之搭配的是在地上展开的中国地图,表达着保利发展要将产品推广至全国的决心。

只有聚焦眼界,才能度出最合适的比例。只有尊重传统与现实,才能实践出最真切的人居体验。只有付诸追求极致品质的实践,才能雕琢不随时光褪色的经典。保利地产根据对新一代居住模式的分析,建立了保利地产产品体验中心,诠释了保利地产"和悦系"全生命周期住宅。

Heyue Exhibition Hall

Adhering to the concept of "fellow travelers good life", Poly Development Company has always been dedicated to researching, developing and creating a living space fitting to the concept of urban regeneration. Based on the analysis of new-generation living patterns, a product experience center is built to interpret the life-cycle apartment of "He Yue" of Poly Development.

The He Yue exhibition hall is designed on the basis of the motion line of visitors. "He Yue" residential products are presented in different areas such as product history, product standards, and community construction, etc. A simple showcase is set next to the reception to show the full-cycle layout of the whole apartment, allowing visitors to have a preliminary idea on the space. After walking into the front hall, the development history of products is displayed on the arc-shaped display wall together with a map of China on the ground, showing Poly Development Company's determination of promoting products nationwide.

Only by focusing can we get the best ratio. Only by respecting tradition and reality can we deliver the most authentic residential experience. Only by pursuing the supreme quality can we create ever-lasting classic. Poly Real Estate established the Product Experience Center based on an analysis of the residential habits of the new generation. The Exhibition Hall represents the full-life-cycle residence of Poly's projects of "Heyue" (harmony and happiness).

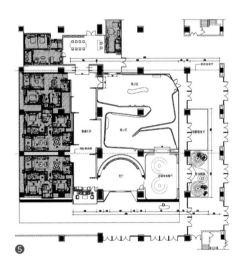

❺ 产品体验中心平面布置图。
Floor plan of the product experience center.

随着动线转入多媒体体验厅,观者可从视觉、听觉、触觉等多方面了解概念的演变与发展,设计师结合各式材料与影音媒体,多方面的展现全生命周期的社区规划与发展,为参观者呈现保利地产对人居美好生活的蓝图。走入展示区,首先引入眼帘的是保利发展的"5P战略",以筑力中国人居梦为目标,打造智能化社区,推动社区中的互动互融。

对于室内空间的规划建设遵从"更强的风格包容性"的设计主旨,保利从解决居住需求出发,在户型布局上强调科学性、合理性,以及更多的复合可能。针对不同阶段的居住需求,在适老化、适幼化及收纳方面建立标准化的体系,满足不同时期家庭对居住空间的变化需要。

Combining various materials and audio-visual media, life-cycle community planning and development have been displayed in multiple aspects, presenting the blueprint of Poly Real Estate to living a better life. The "5P Strategy" of Poly Development Company is the first thing noticed in the exhibition area. The "5P Strategy" is to create an intelligent community and promote the interaction and mutual integration in the community with an aim of assisting to realize the Chinese's dream of owning an apartment.

Based on solving the residence demands, Poly Development Company emphasizes the scientificity, rationality, and more complex possibilities in the layout of apartments while following the design subject of strong style containment in the planning of interior space. Different families' changing needs on living space can be satisfied within the standardized system set up for adapting to the elderly, the child and storage on the basis of living demands at different stages.

1. 全生命周期中强收纳的展示。
   Strong storage in the whole life cycle is displayed.
2. 住宅标准化的展示。
   Display of residential standardization.
3. 保利发展的5P战略。
   The 5P strategy of Poly Development Company.
4. 互联网+板块和个性定制板块。
   Internet + section and personalized customization section.

1. 展现利民服务的内容。
   The contents of convenient services are displayed.
2. 利民服务中比邻洗衣的概念。
   The concept of neighboring laundry in the convenient services.
3. 4. 利民服务中比邻超市的概念。
   The concept of neighboring supermarkets in the convenient services.
5. 安民服务的内容的展示。
   The content of reassuring service is displayed.
7. 舒敏服务内容的展示。
   The content of comforting service is displayed.
6. 8. 9. 健民服务的内容的展示。
   The content of health service is displayed.

在展厅的最后,设计师重点介绍了保利发展对于整个社区的建设构想,秉承着"利民、安民、舒民、和民、健民"的五个宗旨,从不同方面完善社区的基础建设,满足住户日常生活所需。同时,注重社区邻里关系,致力打造和谐社区。从购物、饮水安全,居民娱乐活动及空间规划等方面构建美好的社区生活。整个展厅的设计,没有过多的装饰点缀,设计师用最直接方式向公众展示美好生活的蓝图,完整展现保利发展所规划"美好生活"的模样。

At the end of the exhibition hall, Poly Development Company's construction concept for the whole community has been primarily introduced. the infrastructure of the community will be improved from all aspects to meet residents' daily living demands. Meanwhile, attentions are paid to community neighborhoods with an effort to build a harmonious community. A good community life is constructed from shopping, the safety of drinking water, residential recreation and spatial planning.

❶ 公共办公区域大面积黑色天花的裸露，是管理者所激励员工在黑暗中探索未来的表达。
It's the best era for innovators when information technology enters the era of mobile Internet; when the power in the traditional mode is overturned. Poly Development has continuously optimized the industrial structures for seeking new development opportunities on the basis of keeping a foothold and strengthening in the master business.

❷❸❹ 接待前台处，大面积木色与黑色的混搭激发出别样的火花。
Distinctive sparks are excited with a mixture of large-area original color of woods and black at the reception.

❺ 开放式办公室设计中考究的是开放与独立的关系。
The relationship between openness and independence is considered in the design of open offices.

天马行空的创新办公空间

当信息化进入移动互联的时代，当颠覆传统商业模式的力量风起云涌，这个时代对于勇于创新者而已是一个最好的时代。创新的力量就如无垠宇宙中的点点星光，在无光的黑暗空间中闪耀着光芒。保利发展在立足主业、做强主业的同时，不断优化产业结构，寻求新的发展机遇。

宇宙的神秘，在于无垠的沉默，带给人们无尽的遐想。如同宇宙般的黑色，在本案中获得了淋漓尽致的使用，接待台背景墙，大面积的黑色天花裸露，还原物理肌理的地面。没有一种表达比沉默更有力量，没有一种色彩比黑色来得更让人蠢蠢欲动且受到震撼。

创新是一种想象力，是天马行空的实现。点缀在"宇宙"中的灯饰，如星球般点亮这一片神秘的X地带，创新投资的思维如同探索发现宇宙奥秘。开放式休闲区，红砖材质的墙面搭配半封闭天花和混凝土色的地板，连接着会客区、休息区，通过落地玻璃的透射，明亮的光线从而让整个空间自然、舒适。静坐在休闲吧上欣赏窗外的景色，创意的想法源源迸发。

Unrestrained and innovative office

It's the best era for innovators when information technology enters the era of mobile Internet; when the power in the traditional mode is overturned. Poly Development has continuously optimized the industrial structures for seeking new development opportunities on the basis of keeping a foothold and strengthening in the master business.

The black color like the universe is largely applied in this case, from the background wall of the reception counter, a large area of the ceiling, to the ground. No expression is more powerful than silence just like no color is more stimulating and astonishing than black.

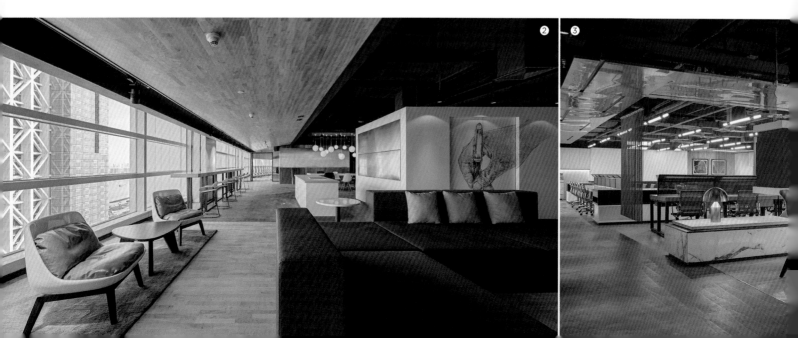

The open leisure area, with red brick walls, semi-enclosed ceilings and concrete-colored floors, connects the reception area and the rest area. The bright light, through the floor-to-ceiling glass, makes the whole space natural and comfortable. Sitting quietly at the lounge, enjoying the scenery from outside, one may come up with innovative ideas.

❶ 休闲区中用跳跃的绿色与橙色营造出轻松活跃的气氛。
Vibrant green and orange sets up a relaxed and vivid atmosphere in the leisure area.
❷❸ 此区域可用作员工休憩或是会客接待。
The area can be a place for employees to relax or to receive guests.
❹❺ 简单的插画和影像等创意元素点缀空间，激发员工无限创意。
The space is decorated with creative elements such as simple illustrations and images to inspire employees' endless creations.

黑的"无色"正是它能够表达蕴含的力量并且与他色搭配的特质。对黑色的敬畏与挑战，渴望追求着一种内在精神力量的彰显与爆发。以象征自然系的木色，搭配少许工业感的现代简约风格，给人自然质朴的感受。半封闭的天花和水泥色般的地板相互带出些许工业感，搭配木色、橙色办公家具，个性化的构思，为办公者增添了一份动力展现。

会议室空间宽敞，两边江景映入眼帘，空间色调以灰色和白色为主，再融入鲜绿色的清新和木色的自然。配合大厦整体地送鲜风系统，给人眼前一亮的感觉，充满朝气与活力。

在仿若浩瀚星空的设计中，保利（横琴）创新产业投资管理有限公司办公室展现出了积聚的力量，创新的工作及休闲模式。黑色中蕴藏的无限可能，表达了保利发展内在的沉稳与对创意无线的追求，跳跃的色系搭配，爆发的是创新的Idea。

The "colorless" black allows it to express the potential power and fit other colors. The office applies natural wooden color coupled with modern minimalist style with some industrial sense, delivering a natural and simple air. Semi-enclosed ceilings and cement-like floors bring a touch of industrial sense to each other. With wooden and orange office furniture, and the application of individualized concepts, such vibes give staff power and energy.

The conference rooms are spacious with river views on both sides. Gray and white as the main tone, a touch of fresh green and wooden color, and fresh air system in the whole building... all of these bring to people a refreshing feeling, full of vigor and vitality.

In the design of simulating the vast starry sky, the office of Poly (Hengqin) Innovation Industry Investment Management Co., Ltd. shows accumulated power as well as innovative work and leisure modes. There are infinite possibilities in black, which shows the inner stability of Poly Development and the pursuit of creative ideas. Innovative ideas can be exploded through vibrantly matching colors.

❶❷ 纵览城市风光，在此构想天马行空的梦。
An unrestrained dream can be visualized here with an overview of urban scenery.
❸ 清新的会议空间。
Fresh meeting space.
❹❺❻ 创新型公共办公空间。
Innovative public office space.
❼ 办公室平面布局手绘图。
Freehand sketching of office plan layout.

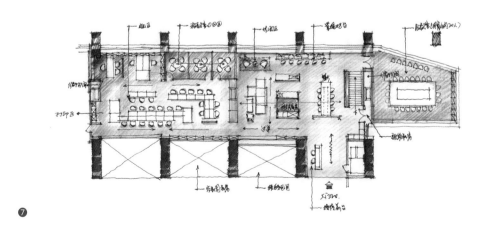

开放互动的办公空间

规整和错落，两个看似矛盾的名词，通过巧妙的空间设计擦出了别样的火花。本案对空间重新建构，清晰的分区采用开放或半开放的设计，模糊了生硬的边界却又保留了秩序，满足人与人之间的办公、社交需要。

穿梭在开放式办公空间以拉近员工之间的距离、建立员工之间良好的社交关系，更好地促进工作的进展与完成。通过灯光的延伸，实现空间视觉的延伸。下藏式灯光将会议室以及办公区域进行区分，同时带来视觉的冲击。不规则的U形讨论区的布置更强调自由、随性，既满足了会议需要，同时适用于日常交流。

用天马行空的想象力，把水波涟漪装进空间里的半开放式的书吧，用错落布置的家具营造出空间的不同利用方式，或办公，或小型会议，或安静地读一本书。灵活布置的显得随性，实际上也具备了调整"空间角色"的功能，尽显巧思。明亮的色彩无疑成为了秩序和灵动的桥梁，恰如其分地增添了空气中的活跃感，加上灯光对空间感的延伸，让人穿梭在每个角落，都能迅速打开头脑，享受一切灵感的冲击。

Inspiring space

Neat and irregular, two seemingly contradictory nouns, light a different kind of spark through clever interior designs. The case adopts an open or semi-open design in reconstruction and clear division of the space, which blurs the rigid boundary but maintains the order and meets the office and social needs.

The open office benefits the bonding between employees and promotes the work progress. The light extension brings the extension of space vision. Such visual impact is achieved by the hidden lighting which divides the meeting room and the office area. The layout of the irregular U-shaped discussion area emphasizes freedom and casualness, which meets the needs of conference and daily communication.

Let the unrestrained imagination run wild, putting waves and ripples into the semi-open book bar and creating different space functions (such as working, holding a small meeting or reading a book quietly) with cleverly arranged furniture. Flexible layout seems to be casual, but in fact helps to adjust the "space role". The bright colors undoubtedly serve as the bridge of order and flexibility, adding an air of vitality. With the extension of the space through light, people can quickly open up their minds and embrace all sorts of inspirations.

❶ 接待前区如一条河流，接待台为白色的帆船，船头点一盏灯指引着前行的方向。
The reception area is just like a river; of which, the reception desk is a sailing boat in white; and a lamp turned on is placed on the bow of the boat, guiding the advancing direction of the boat.
❷ 接待前区天花的设计如在河面上泛起的涟漪。
The ceiling at the reception is designed like riffles on the river.
❸ 设计师用半透明的灰色玻璃，让会议空间不再紧闭。
The meeting area is no more compact using the translucent gray glass.
❹ 办公室彩色平面布置图。
Colorful floor plan of offices.

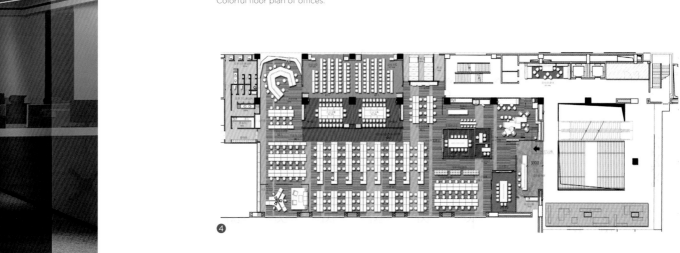

❹

❶❸ 休闲区中明亮的色彩与不同图案的碰撞。
Bright colors are collided with different patterns in the recreation area.
❷ 摒弃单调的办公设计，设计师用简单的色彩搭配与图片的装点，形成独具一格的个性空间。
Getting rid of the monotonous office design, simple color matching and picture decoration are adopted by the designer to form a unique personality space.
❹❺ 办公空间中随意可见到创意的图形，如公共办公区域蜂巢吊灯。
Creative graphics can be seen in the office space, such as the honeycomb ceiling lamp in public office area.
❻❼ 休闲区内组合型沙发，保障个人享有相对独立的空间。
The combined sofa set in the leisure area allows users to enjoy a relatively independent space.

1. 屏风画为北宋王希孟创作《千里江山图》。
   The screen painting is the "Vast Land" created by Wang Ximeng of the Northern Song Dynasty.
2. 打开屏风后可见，一线江景。
   The best river view can be seen after opening the screen.
3. 接待中心的平面布置图。
   Floor plan of the VIP center.
4. VIP接待区软装配饰主要以金属材料为主。
   The soft decoration in the VIP center is dominated by metal materials.

贵宾接待中心

贵宾接待中心朝向珠江，拥揽一线珠江景色。设计师以此为设计构想，将祖国的山水元素融入到设计当中。接待中心主要选用大理石及木质材料为主，软装大多选用金色配饰，打造奢华场所。VIP接待区正向珠江，因此设计师将《千里江山图》喷绘在自动屏风上，与窗外的城市景色相呼应。将自动屏风放在此处，满足不同场景的功能切换。企业展厅的设计开敞明亮，展览形式新颖、现代而不失经典。设计师将展厅设置为开放式的空间，与历史长廊连通，企业的文化与历史一脉相连。展厅中央以一张琶洲新区的摄影图为背景，展现出城市山河美景的同时亦表达企业对城市规划的使命感。

VIP reception center

Facing the Pearl River, the VIP center can enjoy the best view of the Pearl River. With this as a design concept, marbles and wood materials are primarily selected. Meanwhile, golden accessories are mainly applied in the soft decoration to create a luxurious venue. Also, the painting of Vast Land is printed on the automatic screen for echoing with the urban scenery outside of the window. The open and bright design of the company exhibition hall is novel in display form, featuring both modern and classic. The exhibition hall set as an open space is connected to the historical corridor, as the culture and the history of the company are intricately linked. A picture of the Pazhou New Zone is set as the background at the center of the exhibition hall, which can not only show the beauty of urban nature, but also expresses the company's mission of urban planning.

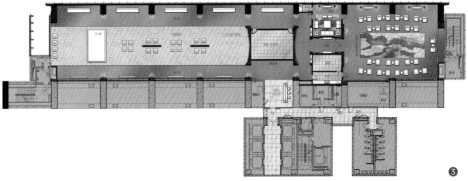

❶ 以琶洲新区手绘图为背景的企业展厅。
The corporate exhibition hall with the freehand sketching of Pazhou New Zone as the background.

❷ 项目的模型展示区。
Model displays of the project.

❸ 企业发展历史走廊。
Enterprise Development History Corridor.

广州保利天幕广场
POLY SKYLINE PLAZA, GUANGZHOU

关键词：一线江景、多元空间、智能化
Keywords: full view of the river, multi-dimensional space, intelligence

建筑设计 / 美国SOM建筑事务所
景观设计 / 美国SWA公司
建筑面积 / 310,000 平方米
委托范围 / 硬装及软装设计
委托面积 / 4,792 平方米

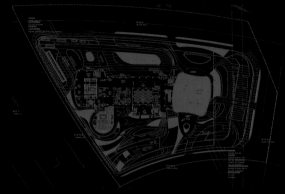

城市的美艳繁华，不在高处，又如何赏识它的美丽。琶洲，从繁荣的对外港口到如今高楼林立，琶洲新城一直是广州与国际对话交流的窗口，吸引无数的海外投资者在此停驻。珠江两岸，一线江边，林立着幢幢建筑，代表着广州名片。现今琶洲地域坐落一处311米高的新坐标——广州保利天幕广场，其顶部楼层与广州城和珠江呈双"弓"型，人们形象称为"琶洲之眼"，为广州这座国际大都市的天际线增色了不少。

项目坐落于珠江边上，270°天际江景，推窗即见广州城。311米超高层写字楼的冠顶会所，拥有内凹曲线外立面，东望粤江入海，西向城际CBD，北临国际金融城，南接生态湿地园，观城市的日月星辰，赏山河美景。在此，居高临下，俯瞰全城灯火，尽览满江烟雨，坐谈资本运作。

The beautiful part of the city can be appreciated at a high point. Pazhou is changed from a prosperous foreign port to an area with high-rise buildings. Pazhou New Zone has always been a window of communicating between Guangzhou and the world, attracting countless overseas investors. At present, Pazhou is located in Poly Skyline Plaza, a new 311-meter-high landmark. Its top floor is presented in double-bowed with Guangzhou City and Pearl River, which is called "Eye of Pazhou", enhancing the charm of the skyline of Guangzhou.

Locating at the bank of the Pearl River, the project enjoys a skyline river view of 270°. The Guangzhou downtown can be seen after opening the window. Guanding Clubhouse locating at the 311-square-meter super high-rise office building has a concave and curvilinear facade, which is east to the Yuejiang River, west to the CBD, north to the International Financial City. It's really an ideal place to observe the sun, the moon and the stars as well as to appreciate beautiful views of landscapes. You can overlook the city lights and watch rivers while discussing capital operation by looking down from such a height.

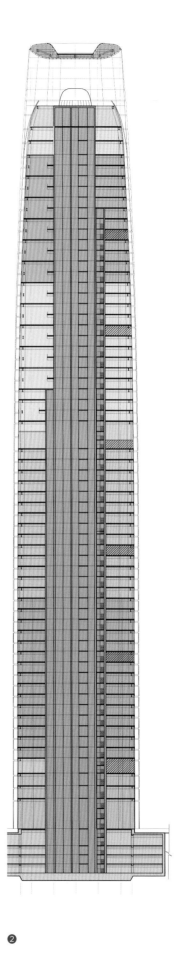

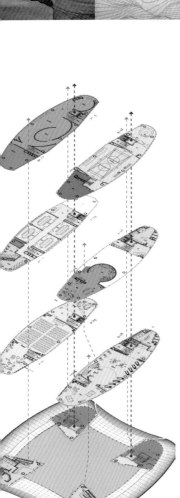

❶ 办公塔楼建筑外观效果图。
Appearance rendering of the office tower building.
❷ 建筑立面图。
Building elevation.
❸ 设计灵感素材：江面，曲线，峡谷和地势线条。
Materials inspiring design: include river surface, curve, canyon and terrain line.
❹ 当代艺术家印象中的中国山水。
Chinese landscapes in contemporary artists' impressions.
❺ "企业文化中心"的功能分布概念图。
Concept plan of functional distribution of "Corporate Cultural Center".
❻ 琶洲鸟瞰图。
Aerial view of Pazhou.

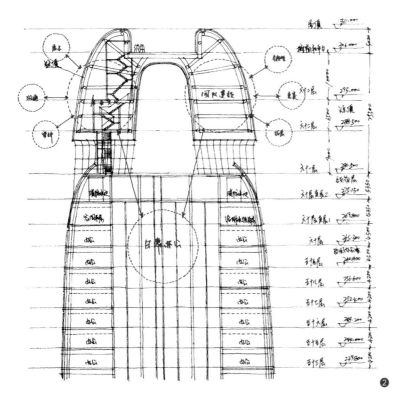

① 办公塔楼-塔楼顶部剖面。
Office tower-section of the top floor of the tower.
② 顶层会所功能分布手绘图。
Freehand sketching of function distributions of the clubhouse on the top floor.
③ 东塔竖剖图，建筑与室内亮度关系。
Vertical section of the east tower, the relationship between the building and the indoor brightness.
④ 从西塔竖剖图可以看到，储藏室相对于62层的夹层储藏空间更亮。
It can be seen from the vertical view of the West Tower that the storage room is brighter than the mezzanine storage space on the 62nd floor.
⑤ 63层的卫生间靠近建筑顶部的中空位置，所以灯光亮度上更靠近室外的灯光。
The bathroom on the 63rd floor is near the hollow position at the top of the building, which is close to the lighting outdoor in terms of brightness.
⑥ 东西塔北侧横剖图。
Cross section of the north side of the east and west towers.

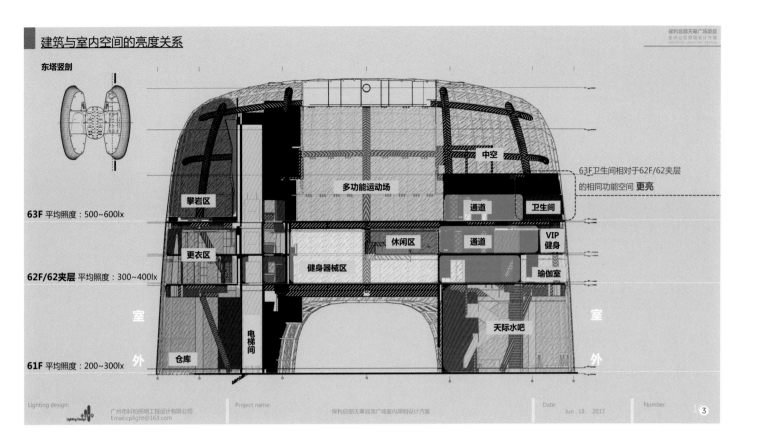

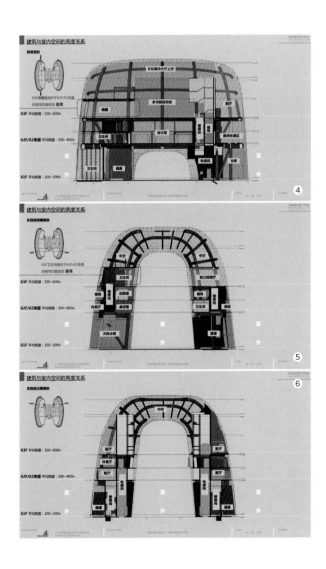

对于顶层会所的设计，设计师从功能角度入手，对东西两座大楼做了合理的功能布局。西塔楼以展示企业文化为主，在展示企业文化的同时也为企业员工提供一个沟通和交流的场所。东侧的塔楼主要用做团队建设，团结企业成员。

我们对内部的灯光也做了一些调节。针对不同空间的分布，不同楼层及区域的灯光，设计师以不同角度的建筑剖面为基础分析，根据空间所在区域的光线视角调节亮度，达到外部光线的视觉统一。

The club at the top floor is designed from the functional aspect. Reasonable function layouts have been made for the east and west buildings. The west tower is primarily used for displaying corporate culture, which is also a place for employees to communicate. The east tower is mainly used for team building and uniting corporate members.

Also, some adjustments are also made on internal lighting. With building profiles at different angles as basic analysis, the brightness of the light distributed in different spaces, different floors and different areas is adjusted based on the space area to achieve the visual unity of external light.

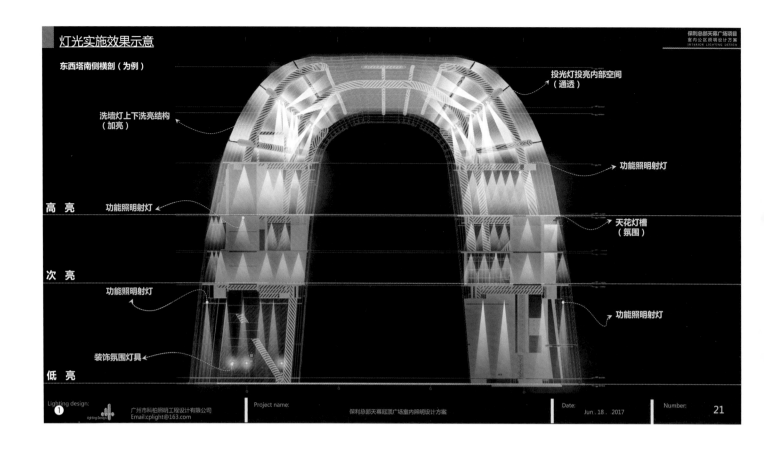

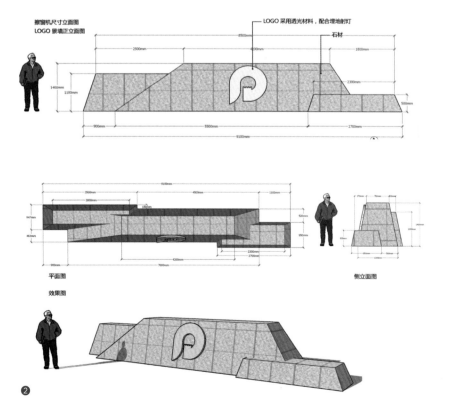

① 灯光实施效果示意图。
Rendering of lighting implementation.
② 特色的logo背景墙的设置，根据人体的视觉为墙体的至高点。
The characteristic logo wall is set as the highest point based on the human vision.
③ 双塔间保留了望江的视野。
The view of river is reserved between the two towers.
④ 全透明的户外广场为员工提供了休憩之所。
A rest place for employees is provided in the transparent outdoor plaza.
⑤ 户外广场风向视图。
Wind direction of the outdoor plaza.
⑥⑦ 户外广场可举行不同类型的活动，或用做员工的休憩之所。
Different types of activities can be organized in the outdoor plaza, or used as a resting place for employees.

61层的户外广场，弧形顶及立面采用全透明的建筑轮廓，地面石刻的企业大事记与一旁的琶洲地图相结合，见证企业的成长和这片土地的日渐变幻。宽阔的户外广场充分的考虑了风向的对流和保留望江的视野，保证广场的对流，为每日劳累的工作者提供了休憩交流之所。四面都有功能化装置雕塑阶梯可供休息与交谈，也为企业方提供举行不同类型活动的空间。

The arc-shaped roof and the facade of the outdoor plaza on the 61th floor are built in the fully-transparent architectural outline. The corporate milestones engraved on the ground stone is combined with the Pazhou map nearby, witnessing the growth of the company and the gradually changing land. Convection of wind direction has been fully considered for the board-view outdoor plaza with the river reserved to ensure the convection of the plaza. Also, it's also an ideal place for workers to rest and communicate. Functional sculpture stairs are set in four ends for rest and conversation, where also provides spaces for the company to hold various activities.

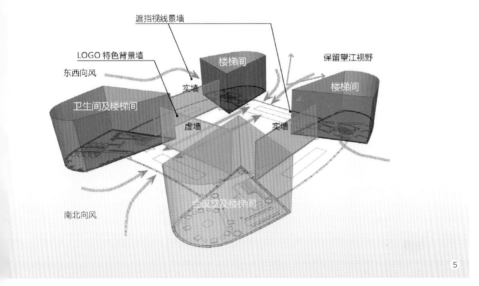

东西塔楼是不同功能的商用空间。西侧的塔楼主要以办公会客为主,在接待客户之余,亦可在此举行会议和商业活动。空间主要以沉稳的灰色为主色调,打造沉稳时尚的商务空间。

东侧则以轻松休闲的功能,62-63层的云端会所,分休闲区与活动健身区域,休闲区室内以大理石材料为基底,木质材料作为辅助搭配,设计师化繁为简,高度保留了建筑结构的体态,用最简单的配饰给予员工舒适之所。活动健身区沿用木饰面材料,颜色选用上更加丰富更具活力的橙色和绿色加入空间,注入了更多的生气和活力。不同功能的健身设备拥有独立的区域,避免了使用时产生冲突。值得一提的是,设计师在此利用充裕的楼高设计飘台,为健身者提供了休憩交流的空间。

The East and West Towers are business spaces with different functions. The towers on the west are mainly offices and meeting rooms, which can be used for receiving guests, holding meetings and commercial events. A calm and stylish business space is created with grey as the dominant tone.

The club on 62-63F consists of leisure area and fitness area. For the former, the designer applies the marble as the main material, decorated with wooden veneer, maintaining the building structure to a large extent and ensuring a comfortable place for workers with the simplest furnishings. For the latter, the designer continues to use wooden veneer material coupled with more diverse and energetic colors. Different functional fitness equipment has a separate area to avoid conflict when used. It is worth mentioning that designers here take advantage of the ample floor height design platform, for fitness people to provide a space for rest and communication.

❶❷ 大理石纹与木饰面的结合,打造现代时尚的商务空间。
Marbling and wood veneer are combined to create a modern and stylish business space.

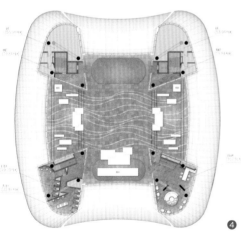
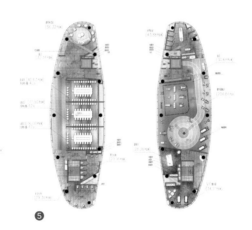

3. 跳跃的橙色为空间注入无限的活力。
   Infinite vitality is injected into the space with the vibrant color of orange.
4. 61层户外广场平面图。
   Floor plan of the outdoor plaza on the 61st floor.
5. 东西两塔的62层平面图。
   Floor plan of the 62nd floor of the east and west towers.
6. 健身区。
   Fitness area.
7. 攀岩及休憩区域。
   Rock climbing and rest area.
8. 商务洽谈区域。
   Business discussion area.

63层的互动展厅是冠顶会所的亮点，设计师摒弃了固有展厅的展示形式，以互动的形式为来访者提供不一样的体验感。整个展厅给人感觉如同置身于河流之中，流线型的设计，引领人们随着线条的方向深入展厅，全展厅的智能化控制系统，令体验变得更加独具一格。如同碗状的装置，实则是可触的电子屏，展示着企业的历史事迹，顶上的铜盘吊饰是可变动的电子可控装置，依据需求变幻不同的形状。展示墙内，远看似乎与一般的无异，但走近一些，便有动画显示在墙面上实现近距离的互动展示。

顶层会所使用大面积玻璃幕墙，让室内外景色相通相连，天空似乎触手可及，四方美尽收眼底，室内主要选用大理石及木饰面来装饰打造休闲会所，材料的色泽在阳光的投射下，更显干练与高贵；智能系统的加入，令展示空间变得更加灵动多变，体验感变强。

❶ 如碗状的电子屏和智能化的铜盘吊饰。
A bowl-shaped electronic screen and intelligent copper plate pendant.
❷❸ 流线型的展厅设计，引领观者顺其深入展厅内部。
Visitors attracted by the streamlined design are guided to the interior of the exhibition hall.
❹ 设计师从地形的线条提取灵感将纹理化作灯饰点亮历史展厅。
Inspired by the terrain line, textures are used for lighting to light up the historical exhibition hall.
❺ 企业展厅的平面图。
Floor plan of the corporate exhibition hall.
❻ 展厅设计的灵感元素。
Inspirational elements in the design of the exhibition hall.

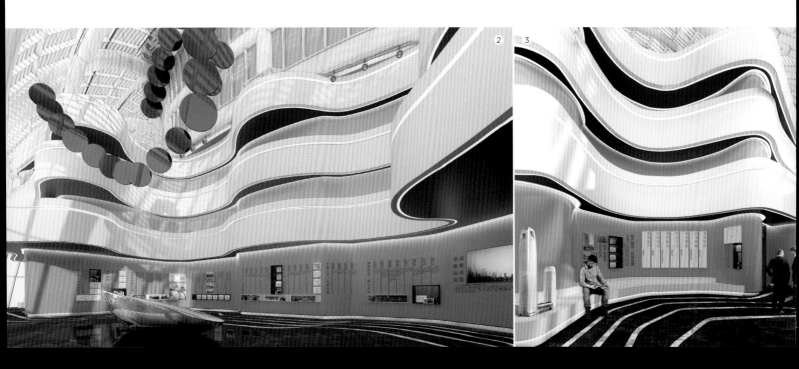

The interactive exhibition hall on the 63rd floor is the highlight of the Guanding Club. People attracted by the streamlined design are guided to the exhibition hall along with the line direction. Moreover, the visiting experience is unique with the intelligent control system arranged in the whole exhibition hall. Although it's a bowl-shaped device, it's actually a touchable electronic screen showing the history of the company. The pendant of copper plate charm on the top is a variable electronically-controllable device that can be changed in different shape as needed.

A large-area glass curtain wall is adopted on the club on the top floor, so that indoor and outdoor scenery can be connected. Meanwhile, the sky seems to be within reach with beauties on the four sides. The interior is primarily decorated with marbling and wood veneer to create a recreation club. Also, material colors are clear-cut and noble in the sunshine. Moreover, the display space is flexible and changeable with the intelligent system, making a strong sense of experience.

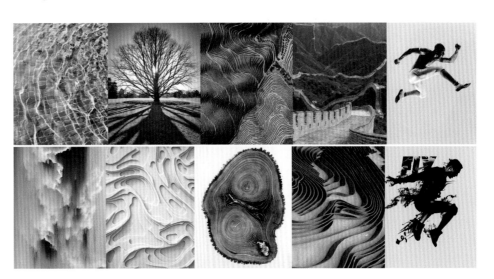

广州华南国际港航服务中心
INTERNATIONAL PORT AND WATERWAY SERVICE CENTER, GUANGZHOU

关键词：肌理、商务地标、滨海港口
Keywords: texture, business landmark, port

建筑设计 / 德国ISA意厦国际设计集团
景观设计 / 三泽园林
建筑面积 / 140,443.5 平方米
委托范围 / 硬装及软装设计
委托面积 / 1,828 平方米

广州港集团自1982年成立，业务涉及港口运输、地产开发、水产养殖、商旅金融等板块。其中地产开发板块以广州市打造"珠江黄金岸线"和黄埔区实施临港经济区规划为契机，推进旧码头转型开发。由其开发运营的华南国际港航服务中心项目被列为2013年广州八大重点改造项目之一，期待打造华南航运服务CBD，并通过5至10年努力，实现与广州中心城区的无缝对接。

作为广州东首席商务地标，华南国际港航服务中心毗邻繁忙的港口，熙熙攘攘的贸易景象自古到今都未改变，于是我们从港口大量的集装箱提取线条和肌理，将现代的时尚感与历史沧桑感完美融合，体现内部空间多维度的层次。

Guangzhou Port Group specializes in port transportation, real estate development, aquaculture, business and finance, etc. The South China International Port Service Center project developed and operated by Guangzhou Port Group is listed as one of the eight key renovation projects in Guangzhou in 2013. It is expected to build a CBD for the South China Shipping Services and to transform Huangpu District.

As the No.1 business landmark in the east of Guangzhou, Southern China International Ports and Airlines Service Center has always witnessed busy trading. At the inspiration of containers, we use their lines and texture coupled with the combination of modern fashion and vicissitudes of history, presenting multi-dimensional layers of space.

新旧时空交汇之处

旋转门开启,时空仿佛交汇于此。"有朋自远方来,不亦乐乎?"世界友人在此能尽情感受广州城市的新旧交融、开放包容以及活力友善。设计选用纹理干净利落的材质以及黑白灰金四色,体现时尚高端的商务氛围。

踏入入户大堂,立刻映入眼帘的是挑高的大理石面墙。墙身的纹理笔直流畅、凹凸有致、简约大气,那灵感正是来源于集装箱箱体的肌理,它是贯穿整幢大楼室内设计风格的基因,也与建筑外观保持一致,遥相呼应。环顾四周,阵列的柱子巨大如机械吊臂、工业纹路让人联想到港口边的桥式起重机。在条条竖线的导视下,人们缓缓仰望,从刻意营造的箱式设计中感受震撼的空间张力。

休憩区的座椅化为一幅世界地图,上面标记了世界著名港口的位置和中英文名称,耀眼夺目,既呈现了集团的企业历程,又散发了广州这座古老港口城市开放包容的魅力和气质。接待处的水吧,外观如同一座船坞。参观游览完毕,客人可到接待处一享茶歇,消磨时光。大楼安装了多台根据功能规划、楼层用途、人数容量等情景配置的先进电梯,保证水平和竖直两个维度的畅通。

大楼安装了多台根据功能规划、楼层用途、人数容量等情景配置的先进电梯,保证水平和竖直两个维度的畅通。

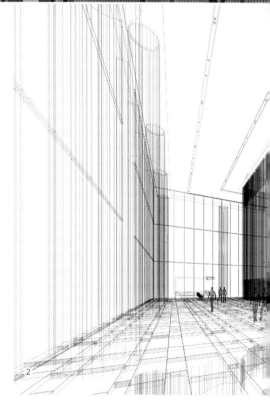

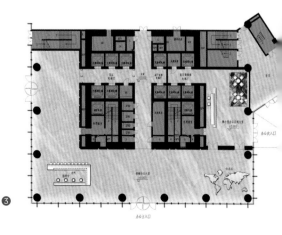

❶ 塔楼办公大堂的主体线条清晰有力,体现出空间的大气端正。
   The clear and powerful line of the main part of the office hall of the tower present a decent and upright space.
❷ 大堂概念手绘图。
   Concept freehand sketching of the lobby.
❸ 图为塔楼办公大堂平面图。
   Floor plan of the office lobby of the tower is shown in the figure.
❹ 阳光由落地玻璃洒向休憩区,世界地图样式的座椅是点睛之笔。
   Sunshine spread to the rest area via the floor-to-ceiling glass, while the chairs in the world map style a finish touch.
❺ 图为微型办公大堂的入口。根据楼层划分电梯区间,有效引导访客分流。
   The entrance to the micro office lobby is shown in the figure. The elevator section is divided according to the floors to effectively separate visitors.

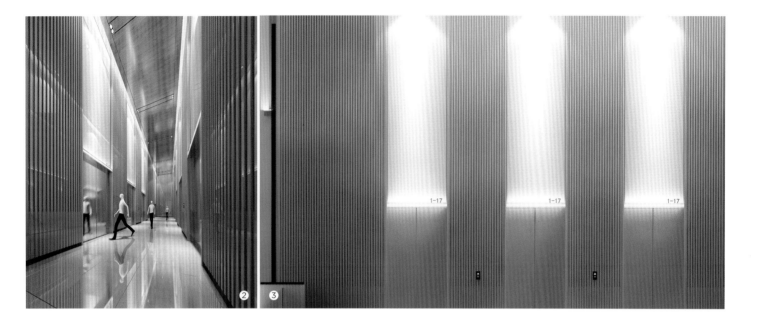

Intersection of old and new times

Friends from different parts of the world come here and rejoice in Guangzhou's fusion of old and new, openness and inclusiveness, vigor and amicability. The designer applies materials with neat texture and black, white, grey, and gold colors, creating a fashionable and high-end business atmosphere.

Walking into the lobby, what comes into sight is a very tall marble face wall with simple and neat texture, which is inspired by containers. And it is also the "gene" that goes through the interior design in the whole building. Looking around, arrays of huge columns remind people of mechanical arms, and industrial texture bridge cranes by the port. Following the lead of vertical lines, people can feel the space tension from the deliberate container-styled design.

Seats at the leisure area are arranged as a world map, on which world-famous ports are marked in Chinese and English. In this way, it shows the corporate history as well as the inclusive charm of Guangzhou as an ancient port city. The bar at the reception looks like a dock where guests can have a tea break after the visiting tour.

The building is installed multiple advanced elevators which are configured according to the corresponding functions and capacity, ensuring clear vertical and horizontal passages.

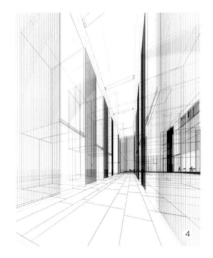

❶❷❸ 大面块之间的对比处理手法也十分有趣。通过材质和灯光语言，光滑与凹凸，平整与韵律得以交互。
Contrast processing methods between large areas is also interesting. Smoothness and concave-convex as well as levelness and rhythm are interacted through materials and lighting languages.
❹ 图为电梯间效果图。
Rendering of the elevator room is shown in the figure.
❺❻ 作为首席地标，中心综合商务、休闲等功能，以服务周边配套。
As the landmark, business and leisure functions can be found in the center to support the surrounding facilities.

光之中庭

建筑内部中分布的三个楼层组（18-29层、30-40层、42-51层）均拥有绝美的中空平台设计，两边仍以"集装箱"为创作灵感的墙体包裹着凹凸竖线，韵律迭起。视觉上的巧妙设计既带动了阳光和空气的流动，也令建筑空间感更强烈。

Bright courtyard

Three floor groups (F18-F29, F30-F40, and F42-F51) in the building enjoy perfect overhead platforms, both sides of which apply concave-convex lines on the wall inspired by "containers". Such clever design promotes the flow of light and air, highlighting the building's sense of space.

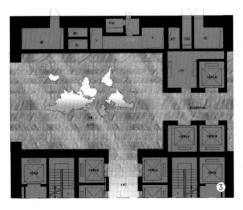
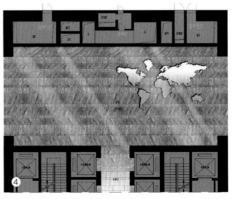
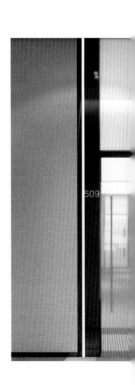

❶❷❺ 为了强调建筑结构和优越的采光条件，中庭的风格仍沿用利落的现代简约，利用光线和线条打造出"光栅"的美妙效果，体现出现代办公商务区高效、灵活的精神内涵。
In order to emphasize the building structure and the excellent lighting conditions, the neat and modern simplicity is still adopted in the atrium. The marvelous effect of "grating" is created with lights and lines, reflecting the efficient and flexible spirit connotation of the modern office business area.
Atrium plan.

❸❹ 中庭平面图。
Floor plan of the atrium.

❻❼❽❾ 根据空间的整体气质，设计主要用木材和石材作为主要的装饰材料，以营造干净纯粹的氛围。
Woods and stones are mainly adopted as main decorative materials based on the overall style of the space, so as to create a clean and pure atmosphere.

❶ 标准办公室既拥有高层风光,也适应当代办公空间的设计潮流。
　The standard offices have high-rise scenery and the design trend of the modern office space.
❷ 阁楼的元素依然有保留且功能会更多元化:集多功能会议室、董事长办公室、商务接待于一体,用色深沉稳重,特显身份地位。
　TThe attic element is reserved for diversified functions: the multifunctional meeting room, the chairman's office, and business reception are integrated with dark color to express stability and unique status.

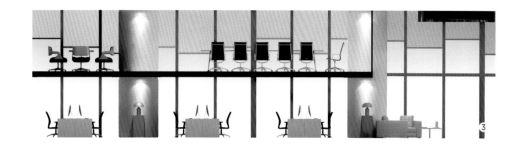

宜业宜居，精英之选

港航服务中心商住两用，主要以年轻化的Loft风格和多样化的户型满足不同人士的需求。麻雀虽小，五脏俱全，复式设计合理地分配了空间，清新的白色和稳重的木色完美融合，雪白柔和的大理石墙体与钢条搭配可谓是刚柔结合。在秩序和美感中拿捏恰到好处的平衡，显示出设计师深厚的功力。

Choice of elites: work-friendly and living-friendly

Ports and Airlines Service Center, for commercial and residential use, adopts Loft style and many other house types to meet different needs. Small but fully-equipped, the property applies duplex designs to reasonably allocate space. The fusion of white and wood color, and the matching of white marble wall and steels, both show the perfect combination of softness and hardness. The perfect balance between order and aesthetics reflects the designer's profound skills.

❸ 办公空间立面图。
Elevation of office spaces.

❹❺ 世界著名的港口城市如鹿特丹、神户、纽约、温哥华等，均深刻地受港口开放融合的文化影响。员工办公区和董事办公室的地毯图案选用华南国际港航服务中心的位置区位地图，寓意着扎根本土，走向国际的当代经营理念。
World-famous port cities such as Rotterdam, Kobe, New York, and Vancouver, etc. are deeply affected the port culture of openness and integration. Regional district map of South China International Port and Shipping Service Center is adopted as carpet patterns of ordinary offices and the director's offices, symbolizing the contemporary business concept of rooting in China and going global.

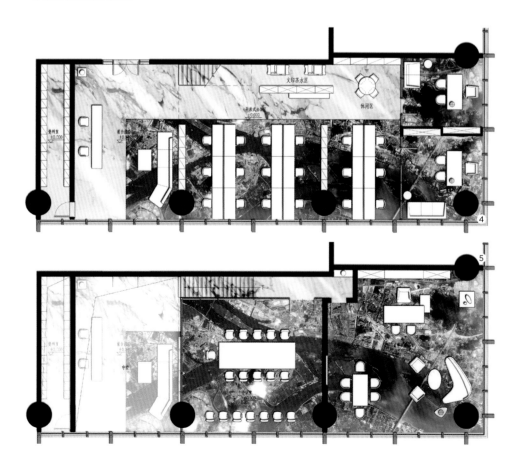

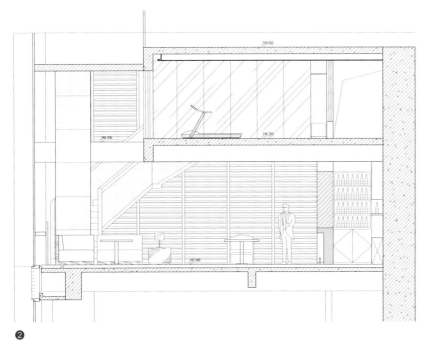

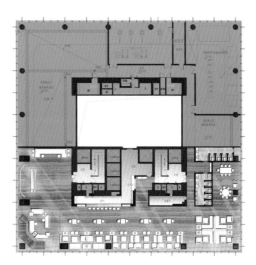

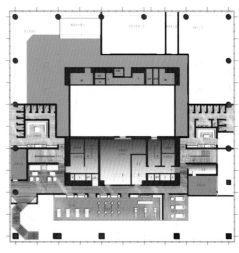

❶ 接待台一扇英伦玉石屏风、典雅柔和；右边的壁画刻画着舞动的五线谱，却又在此刻变成凝固的艺术。
The reception desk is elegant with a British jade screen; the mural painting on the right with the dancing staves becomes a solid art at this moment.

❷ 会所剖面图。
Architectural section of the club.

❸❹ 会所首层、二层平面图。
Floor plan of the first floor and the second floor of the clubhouse.

❺ 会所餐厅酒吧中，百叶窗不但提供了保障了隐私，带来美妙的光影效果。光透过百叶窗，给装修极简的酒吧和洽谈区增添条条饰纹，别有一番风味。
Blinds not only provide not only privacy protection, but also marvelous effects of light and shadow in the restaurant and bar of the clubhouse. Stripe patterns are added in the bar where is decorated simply and the discussion area as light passes through the blinds.

❻❼ 二层为健身区，健身房和瑜伽室分别为好动和好静的来宾提供选择。
The fitness area is on the 2nd floor. The gym and yoga room are provided for guests.

顶层风光，与运动相伴

顶层会所坐拥广州市环回美景，空间环境和谐悦目，弥漫时尚生活的雅致和匠心考究，为商户及访客提供了一个商务休闲、运动健身的好去处。行政酒廊、会议室、健身中心、顶层天际游泳池一应俱全。这里奢华、宁静、美观，如同一个和谐而隐秘的世界，为舒适概念增添一种新的卓越审美内涵。

Views and fitness at the top

The club at the top floor, with a full view of the city and delicate designs, is a good place for business, leisure and fitness. Luxury, peace, and beauty make up this harmonious and secret place, adding a new aesthetic touch to "comfort".

珠海中冶盛世国际广场
MCC HONOUR PLAZA, ZHUHAI

关键词：弧形、互通桥梁
Keywords: arc-shaped, interconnected bridge

建筑设计 / Aedas凯达环球设计公司
景观设计 / ACLA
建筑面积 / 834,908.84 平方米
委托范围 / 硬装设计
委托面积 / 2,381.83 平方米

千岛之市——珠海，毗邻港澳，正以蓬勃之势向上发展，横琴新区借改革的冉冉之火不断前进。借薪火之势，中冶集团在此建立一个集商业、住宅、文娱等多功能为一体大型综合项目——中冶盛世国际广场。广场屹立于珠海横琴口岸服务区之上，隔河对望澳门奢华酒店群，其多类型业态的聚集为横琴商区创造无限商机。289米城市观光厅，金珠浪漫会所等亮点组合，两城美景净收眼底。盛世国际广场坐立于两地的重要交通枢纽，地上地下连廊直通横琴口岸联检大楼，进一步促进两岸文化经济交流。

项目地处毗邻海旁，设计师以锦鲤游动的姿态构想建筑外观，将对该项目的美好祝愿化作建筑符号。建筑主要分为南北两塔，北塔主要是办公功能为主，南塔分为商业及住宅两种，以南北连体的十八层为分隔。南北塔楼的相连，如两条锦鲤的相遇，意珠海与澳门的相互合作，弧形的建筑体在此地更为夺人眼球。

Zhuhai, a city of Shidao, is adjacent to Hong Kong and Macao, which is developed in a vibrant manner. The Hengqin New Area will be always advancing by the advantage of reforming. Taking advantage of the torch, the Shengshi International Plaza of China Metallurgy, a large-scale comprehensive project integrating business, residence, culture and entertainment activities has been established by China Metallurgical Group. Shengshi International Plaza is located at an important transportation hub of the two places. The above-and below-ground corridor directly connecting the Hengqin Port Joint Inspection Complex Building has been built for promoting cross-strait cultural and economic exchanges.

The arc-shaped building is eye-catching, which is composed of north and south towers, the north tower is mainly office function, and the south towers; while the south tower is divided into commercial area and residential area. The connection of the north and south towers, like two fancy carps coming across, indicates the mutual cooperation between Zhuhai and Macao.

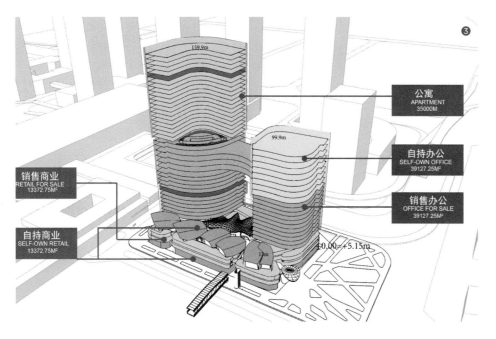
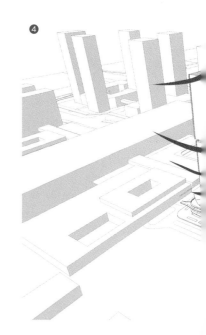

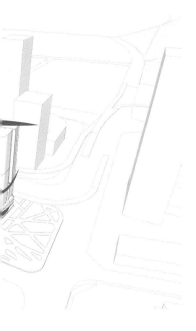

❶❷ 图为中冶口岸大厦建筑外观。
The picture is the appearance of the MCC Port Building.
❸ 图为建筑功能分布。
The picture is the distribution of building functions.
❹ 弧形的建筑体，似锦鲤游动的姿态。
The arc-shaped building is like the swimming posture of fancy carp.

南北互通，文化交融

深处对岸的澳门，深受西方欧式风格的影响，室内装修华丽高贵；而海岸城市珠海，面朝大海，春暖花开。设计师将两种元素都融入空间当中，当澳门建筑的华丽风与海岸城市的蓝色浪相遇，便形成一个南塔的首层大堂。

North-South connectivity and cultural integration

Macao, on the other side of the sea, is deeply influenced by Western European style with gorgeous interior designs; while the coastal city of Zhuhai all-inclusive and versatile. The designer combines the gorgeous Macao style and the blue style of coastal city in the first floor lobby of the south tower.

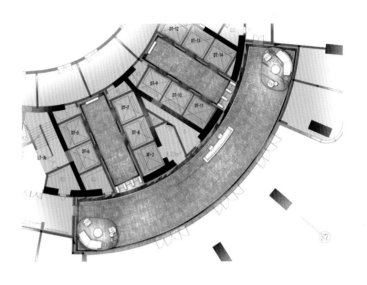

推门而入，南塔的首层大堂如开阔的圆弧展厅，海纳百川，迎接各路客人。一圈圈刻画的圆弧，顺应着建筑原结构在天花展开。波浪式的吊灯，随着线条排开，与正中心的蓝色大理石板相衬，点亮着空间的蓝色的秘密。两侧精致现代家具，为来访客人提供休憩之处，简单的配饰让空间更加宽阔明亮。

Pushing the door open, the first floor lobby of the south tower is like an open arc-shaped exhibition hall, welcoming guests from all walks of life. The arc depicted in circles conforms to the original structure of the building in the ceiling. The wave-like chandeliers, lined up with the blue marble slab in the center, illuminates the space. Exquisite modern furniture on both sides provides a place for visitors to relax, and simple decorations make the lobby wider and brighter.

❶❷ 南塔首层大堂效果图。
Rending of the lobby on the first floor of the south tower.
❸ 南塔首层电梯间。
Elevator room on the first floor of the south tower.
❹ 南塔首层平面分布图。
Floor plan of the first floor of the south tower.

若南塔大堂的装潢如蓝色静谧的大海，北塔则如温润的大地。浅灰色的大理石从大堂延伸至电梯厅，设计师巧妙的线性元素加入其中，像是勾勒着城市的肌理。电梯厅用意大利玉石搭配金色的钢材，钢材的互相交错，像是两岸互通的网路，紧密相连；在软装配饰上，设计师去掉浮夸的装饰，用简单的空间语言打造低调奢华的空间。

十八层的中央连廊，弧形的设计似摆动的鱼尾，金属的标识顺着动线贯穿南北两塔。南北两塔的互通如同珠澳两城，中冶盛世国际广场的建立让原有的交流互动变得更加紧密。大型多功能综合体屹立于横琴口岸中，必将为两岸的发展增添无限商机。

If decorations in the lobby of the south tower is like the blue sea, the decorations in the lobby of the north tower is like the warm land. The light grey marbles are extended from the lobby to the elevator hall. Linear elements are added tactfully, just like outlining the city's textures. The elevator hall is made of Italian jade and golden steel. The intertwined steels are like tightly connected networks of two sides. Getting rid of exaggerated decorations, a low-key luxury space is created with a simple space language.

The eight-story central corridor is designed in an arc shape, being similar to the swinging fishtail. And the metal logos are penetrated through the south and the north towers along with the motion line. The interconnection between the south-north towers is like Zhuhai and Macao. The communication and interaction become closer with the establishment of the Shengshi International Plaza of China Metallurgy. The large-scale multifunctional complex locating in the Hengqin port will definitely add infinite business opportunities to the development of both sides.

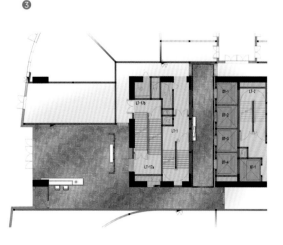

❶❷ 图为北塔首层大堂。
The lobby on the first floor of the north tower is shown in the figure.
❸ 北塔首层平面分布图。
Floor plan of the first floor of the north tower.

④ 意大利玉石与古铜钢搭配打造轻奢电梯间。
A entry luxury elevator room is created with Italian stones and bronzed steels.

⑤ 十八层连廊。
Corridor on the 18th floor.

⑥ 十八层平面图。
Plan of the 18th floor.

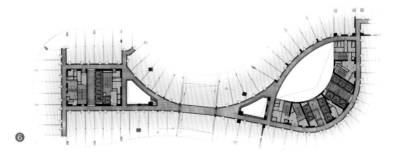

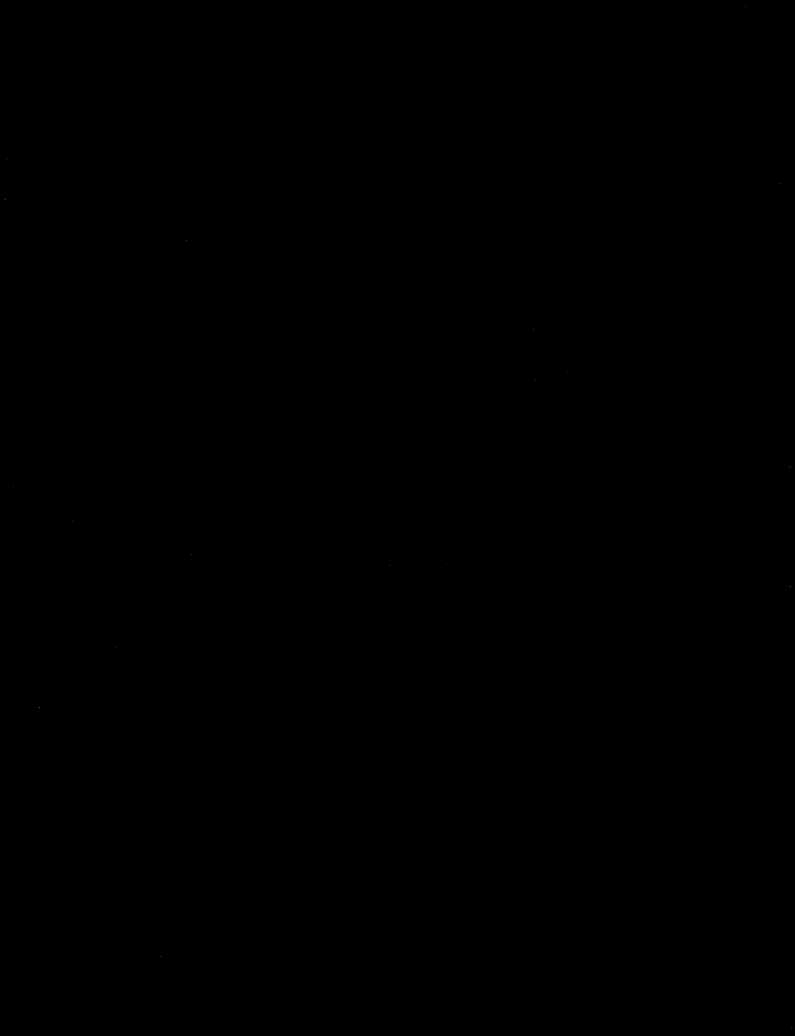

商业空间
COMMERCIAL SPACE

商业空间不仅仅是单纯的商业环境。我们根据商业特性，为各类商业空间合理划分了功能区间，综合性规划周到的设施和视觉元素，为商业环境带来舒适、韵律、均衡和艺术性。

Commercial space is more than just a simple business environment. Based on commercial characteristics, we have divided functional areas for various commercial spaces with comprehensive supporting facilities and visual elements, bringing comfort, rhythm, balance and aesthetics to the business environment.

南京 & 上海诺阁雅酒店
南京保利中央公园销售中心
广州保利天汇会所
深圳海上世界沙龙会所
启迪协信杭州科技城展示中心
上海协信集团企业会所
深圳贝骊洛生活美学馆

Neqta Hotel, Nanjing & Shanghai
Sales Center Of Poly Center Park, Nanjing
Club Of Poly Grand Influx, Guangzhou
Art Salon In Sea World, Shenzhen
Exhibiton Center Of Tusincere Science City, Hangzhou
Sincere Corporate Club, Shanghai
Best & Real Living Zone, Shenzhen

南京&上海诺阁雅酒店
NEQTA HOTEL, NANJING & SHANGHAI

关键词：精品酒店、时尚、商旅
Keywords: boutique hotel, fashion, business trip

建筑及景观设计 / 东方建筑设计院
建筑面积 / 80,000 平方米
委托范围 / 硬装及软装设计
南京委托面积 / 2,950 平方米
上海委托面积 / 18,000 平方米

NEQTA（诺阁雅）系由国际五星级酒店费尔蒙酒店和金大地集团强强联袂打造的五星级服务精品酒店，秉承费尔蒙百年服务精髓，提供独一无二奢华酒店式服务。

酒店涉及到多种空间的分区和功能划分，其复杂程度考验着设计的专业性和综合性。从项目早期开始，SNP就参与整体的规划及落地执行，在建筑条件、照明系统、机电设置、标识系统等诸多方面向酒店管理方提供专业意见和帮助，也在双方的合作过程中不断积累项目整体把控的经验和心得。

Jointly built by the international five-star hotel—Fairmont Hotel, and Keyne Group, NEQTA carries the time-honored tradition of Fairmont and offers unique luxurious hotel service.

The hotel covers multiple divisions of space and functions. Therefore such complexity is in need of professional and comprehensive designs. From the early phase of the project, SNP has engaged in the overall planning and its implementation, providing professional opinions and support in many aspects such as construction conditions, lighting system, mechanical and electric setting, identification system, and etc.

❶❹ 酒店大通过夜间的灯光系统变化、柜体的整体可变，实现了从大堂到时尚酒吧的转变。
The hotel lobby can be changed to a stylish bar through changing the lighting system at night and the overall flexibility of cabinets.

❷❸ 电梯间及电梯内部。
Rendering of the elevator room and elevator.

❺ 各空间手绘图。
Sketches of the spaces.

❻❼ 客房效果图。
Rendering of the hotel rooms.

费尔蒙酒店集团契合新生代商旅人士愈加追求工作与生活完美衔接的生活方式需求，以社交中心作为顾客体验的关键所在，用追求完美、年轻新鲜的设计，多变的空间功能，以及热情、个性化服务打造诺阁雅与别不同的品牌标志。

年轻、开放、满足社交是酒店设计的新趋势和新潮流。为了突出精品酒店的品质感，大堂成为着墨甚多的设计重点。配合灯光系统的明暗处理，对比强烈的配色显得大胆时尚。大堂综合会客交谈、餐饮观光、阅读休憩、健身娱乐等功能于一体，空间的复合能力和吸引力得到有效的提升和优化，满足年轻商务人士的不同需求，打造独一无二的"商旅社区"。

Catering to the business travelers' new life style of striking a balance between work and life, Fairmont Hotel Group focuses on "social" as the key of customer experience, adopts perfection-oriented and refreshing designs with versatile space functions, and provides enthusiastic and personalized service, making NEQTA a unique brand.

Being young, open, and social is a new trend of hotel design. In order to highlight the quality of a boutique hotel, the design of the lobby is a key focus. Shading treatment in line with the lighting system and contrasting colors are bold and fashionable tries. The lobby integrates its space functions (meeting guests, catering, sightseeing, reading, leisure, fitness, entertainment all in one) and meets different needs of young business travelers, creating a unique "business trip community".

诺阁雅让旅居生活变得生动和乐趣，提供年轻个性、热情友好的服务。设计从"蜂巢"这一充满甜蜜、互助、繁荣的自然形象找到灵感，将其融入到酒店视觉元素中去，呈现酒店核心文化，突出酒店的品牌价值所在。大堂不再只有Check in & Check out的冰冷程序，而是在友好、明亮的氛围中迎送八方来客。

NEQTA provides personalized and hospitable service, adding a lot of fun to business trips. Inspired by "hive"—sweet, interdependent, and prosperous, the designer applies it to the hotel design, highlighting the core culture and brand value of the hotel. Lobby, not just for routine check-in and check-out, but welcomes guests in a nice and bright atmosphere.

❷

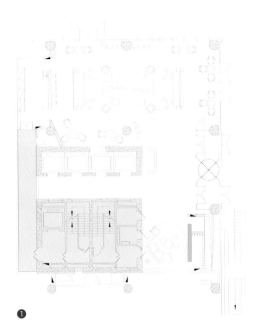

❶

❸

❶ 南京诺阁雅酒店大堂平面图。
Lobby plan of Nanjing NEQTA.
❷❸ 大堂空间手绘图。
Sketches of the lobby.
❹❺❻ "蜂巢"标识系统为精品酒店品牌带来年轻化设计和新的品牌气息。
Young design and a new brand feeling are brought to the boutique hotel brand thanks to the "honeycomb" logo system.

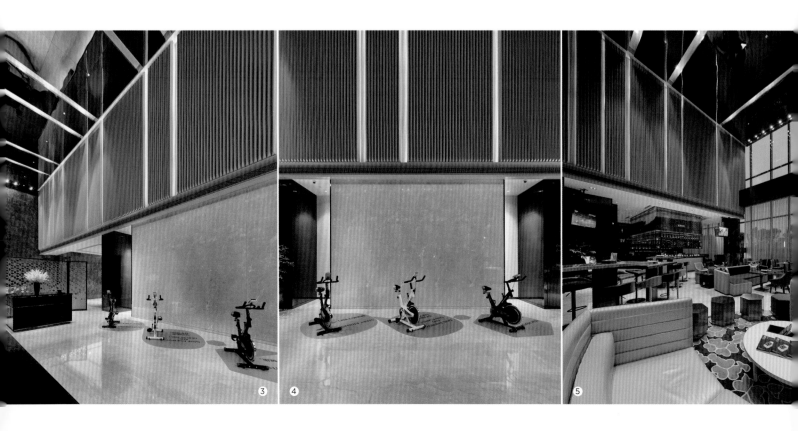

接待前台、小型健身区、会客区沿着走廊过道连接在一起。餐吧区可根据时间调整功能，早上是提供简餐的咖啡区，入夜后则变为小酌一杯的酒吧。造型简约，具有组合潜力的家具为空间营造出友好互动的氛围。

Along the corridor, the reception desk, small fitness zone, and drawing room are connected. The canteen can adjust its functions according to time, for example, it can be a cafeteria offering counter meals in the morning or turn into a bar at night. With simple designs and modular furniture, it creates an amicable interactive atmosphere.

❶ 上海诺阁雅酒店接待前台。
The reception areas in the Shanghai NEQTA.
❷❼❽ 沙发休憩区。
Sofa area.
❸❹ 健身区。
Fitness area.
❺❻ 同样可以变成为夜间酒吧的酒店大堂。
The hotel lobby can be changed to a stylish bar through changing the lighting system at night and the overall flexibility of cabinets.
❾ 大堂平面图（局部）。
Plan of the lobby.

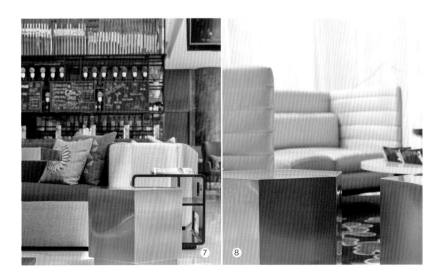

我们基于多年来丰富的住宅户型设计研究和经验，对诺阁雅的客房户型作出全面的提炼和优化，令其更加合理和精准。经过与酒店运营方的协商沟通，设计在满足基本的入住需求外，在原建筑角位、洗手间的综合功能、家具的尺寸及位置等方面都作出了适当的修改和调整，最大程度地提升使用面积和优化行走动线，令各方面符合人体需要和使用友好需求，提升顾客入住的舒适度和归属感，达至宾至如归的效果。

Based on the rich experience on residential design over the years, we have worked out comprehensive quality designs on guest rooms of NEQTA. After the consultation with the hotel operators, besides meeting basic needs, we have made proper adjustments regarding the original building corners, multi-functions of bathrooms, as well as the size and location of furniture, maximizing the usable area and optimizing the flows to the fullest. We aim to be user-friendly in every aspect and increase customers' coziness and sense of belonging during the stay, making them feel like home.

❶❷❸ 酒店客房内部景致。
The interior view of hotel rooms.
❹❺❻❼❽ 房间内的无障碍设计特别照顾到有需要的客人。
The barrier-free design in the room is specially to care for guest in need.
❾ 客房楼层平面图。
Floor plan of guest room.

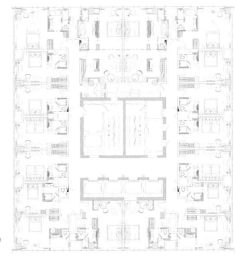

多功能餐厅中灯光及柜体的变化，为创新餐饮概念预留了更多的精彩变化。鲜艳的色彩，富有品质感的设计细节提供丰富且深刻的餐饮体验。值得一提的是，酒店的接待前台不再采用传统的砖石结构，而是相对地更为小巧灵活。前台的柜体可根据使用人数作出体积和功能的变化，接入智能化系统后，旅客可以与酒店工作人员实现更为自由和深入的交流。

The changes of lighting and cabinets at the multi-functional canteen reserve more surprises for creative catering. Bright colors and quality design details offer rich and profound dining experience. It is worth mentioning that the reception desk becomes more exquisite and flexible instead of using traditional stone structure. The size and functions of the desk can be adjusted according to the number of people and with intelligent system a more in-depth and free-style communication could be achieved between customers and staff.

❶❷ 多功能接待前台。
Multi-function reception desk.
❸ 酒店餐厅平面图。
Plan of hotel restaurant.
❹❺❻ 餐厅内部景致。
Inside of the hotel restaurant.
❼ 各空间手绘图。
Sketches of the space.

④

⑤

⑥

⑦

❶❷❸ 空间中灯光的强度、角度也影响入住体验和舒适度。我们综合解决酒店的全灯光系统，有效为后续的室内装饰、视觉表达提供前置方案。恰当、准确的照明设置搭配鲜艳的配色又不觉刺眼，烘托出高端精品酒店的品质感。
The brightness and angles of lighting affects guest experience. With an overall consideration of the hotel lighting system, we come up with a prepositive proposal regarding follow-up interior designs and visual effects. Perfect lighting along with bright and comfortable colors deliver the quality taste of a high-end boutique hotel.

❹ 客房内部手绘图。
Sketch of the guest room.
❺ 客房卫生间。
The bathroom.

南京保利中央公园销售中心
SALES CENTER OF POLY CENTER PARK, NANJING

关键词：工业风、现代优雅、去售楼部化
Keywords: industrial style, modern elegance, de-marketing

建筑及景观设计 / 南京长江都市建筑设计有限公司
建筑面积 / 467,820 平方米
委托范围 / 硬装及软装设计
委托面积 / 555 平方米

在房地产火热的市场背景下，销售中心除了吸引消费者、突出销售门户等硬性功能之外，是否还有其他可能性？作为保利地产第一个情景式售楼部，南京保利中央公园销售中心对我们而言也有里程碑的意义。一个充满想象力的作品，正式融入南京九龙湖核心板块繁华的商圈地带，在完成自身的角色任务之余，也迎来了一次华丽的变身。

空间以化繁为简为主要准则，化身成媲美专业咖啡馆的会谈空间，呈现销售空间精彩的另一面。咖啡令人愉悦的香气弱化了"销售中心"过于强势和刻板的商业气息，客户在休闲轻松的环境中自由交谈，从中加强了对项目本质的体验和感知。

As the first situational sales department of Poly Real Estate, Nanjing Poly Central Park Sales Center is a mile stone. Officially as a part of the Nanjing Jiulong Lake CBD, such imaginative work has done its part and finished a splendid transformation.

Under the principle of being simple, the sales center turns into a drawing room that can compare with a professional cafeteria. The fragrance of coffee softens the tough and rigid commercial sense of a sales center, allowing guests to talk in a relaxing environment and thus gain better experience and understanding of the project.

❶❷❸❹ "去售楼部化"的尝试缓和了过硬的商务气氛，也实现了空间的再利用。
Rigid business atmosphere is eased with an attempt to "remove the style of sales department", which also reuse the space.
❺ 销售中心平面图。
Floor plan of sales center.
❻❼❽ 休闲空间一角。
The corner space of department.

不做作就是工业风的精神

空间风格设计以现代工业气息为主题，人字铺贴法的木地板、格栅天花延伸开去，尽显透视感，又赋予了空间复古优雅的基调。生活的巧思、朴素的质感以及信手拈来的生活趣味是设计师对工业风的精准理解。干净利落且开放式的布局，可容纳多位朋友一同落座轻松交流，共同来营造当代空间的美妙和吸引力。

Industrial style: natural and simple

With modern industrial style as the theme, the space applies "inverted-V-shaped" paving wooden floor and grille ceiling, adding a sense of transparency and vintage elegance to the space. Delicacy, simple texture, and ubiquitous fun are the best interpretation of industrial style. A neat and open layout allows multiple friends to gather and create pleasant moments.

❶❷❸ 沉稳的水泥墙、暖色调的灯光、经典的花砖元素和特意做旧的木饰面。复古装饰风格，带出了工业风独有的历史感，看似老旧却摩登感十足。
Composed concrete wall, warm lighting, tiles with classic flower patterns, and vintage wooden veneer... all of these retro decorations display the unique sense of history of industrial style, vintage but modern.

广州保利天汇会所
CLUB OF POLY GRAND INFLUX, GUANGZHOU

关键词：中央车站、ARTDECO、情景式
Keywords: Central Station, ARTDECO, situational

建筑设计 / 筑博设计股份有限公司
景观设计 / 广州怡境景观设计有限公司
建筑面积 / 1,040,000 平方米
委托范围 / 硬装及软装设计
委托面积：1489 平方米

纽约中央车站诞生于美国最鼎盛的工业时代，又糅合了诸多艺术风格，犹如一座时代的博物馆。它至今仍熙熙攘攘送往迎来，在新与旧、动与静之间给人无限的遐思和灵感。位于广州天河智慧城核心地带的小新塘片区也有创新的力量。延续情景化的设计手法，再度化为精美的"中央车站"，引入皮革、铁艺、火花、红砖石等工业遗迹，利用折现及金属等符号让ArtDeco的富丽堂皇平衡纯工业风的粗粝感。

Built at the US's most prosperous industrial age, New York Central Station is like a museum of times with a mixture of multiple artistic styles. Xiaoxintang Area, located at the heart of the Guangzhou Tianhe Intelligent City, also enjoys creative potential. Continuing to use the situational design, it turns into an exquisite "Central Station", introducing industrial relics like leather, iron art, sparks, and red bricks. The application of broken lines and metals allows the splendidness of ArtDeco to balance the roughness of pure industrial style.

接待前台的背景强调色彩与线条的运用，一匹由钢铁锻造的骏马如点睛之笔赋予空间别样的动感和艺术性。利用建筑本身的层高优势，划分二层作为功能区，预留艺术、教育等软性功能。水吧与洽谈区相连。室内巧克力色的柜体和焦糖色的灯光的组合营造温暖的氛围。

The background of the reception emphases the use of colors and lines. A steed made of steel, a finishing touch of the space, adds special vitality and aesthetics to the room. Leveraging the height advantages of the building, the second floor is served as a functional area reserving for art or educational use.

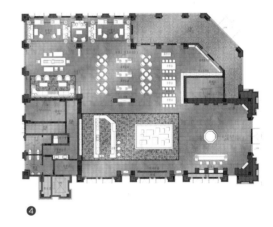

❶❷❸ 根据建筑的体态，内空间利用钢结构模拟中央车站的复古感，使空间与建筑观感达成一致。
The vintage sense of the central station is simulated using the steel structure in the interior space in accordance with the building shape, making consistent space and architectural appearance.

❹ 建筑首层平面图。
Plan of the first floor.

❺ 文创品销售区。
The cultural and creative products sales area.

❻ 水吧区。
The bar.

ArtDeco的优雅无处不在

水吧的背景墙以金属绞盘为装饰，置物层架被细孔丝网包裹，令视觉多了复杂的层次感。"如果有天堂，天堂应该是书店的模样。"文创区成为整个空间最大的亮点和运营延续点。它将作为社区的精神领地，给予社区居民写意的阅读体验和手作的乐趣。

爱迪生灯泡所打造的一系列灯饰散发着温暖迷人的光线。由红砖与黑白插画打造的背景墙，如一扇窗口让人探见车站月台的繁忙。卡座区绿色的皮革沙发座椅温润细腻，复古优雅感扑面而来。

Ubiquitous elegance of ArtDeco

The background of the bar is decorated with metal capstans and the shelves are wrapped by wire nets, visually adding different layers. Cultural and creative area is the highlight of the space, where residents can enjoy reading and handcrafting.

Warm lighting, background wall made up of red bricks and black-and-white illustrations, booth area with green leather sofas... all of these reveal an air of retro elegance.

❶❷ 内部空间掠影。
Pictures of inside room.
❸❹ 金色元素十分迷人。
Golden elements are charming.
❺❻❼ 工业风内饰充满复古的味道。
The industrial interior has a retro feel.

功能性的复合空间，销售中心的社区化

销售中心要融入社区，意味着除视觉美感外，这处空间需要具备更多的实用性。因此设计师为各种社区生活情境和经营业态预留了位置，复合性的功能体系将为空间注入更多活力。

标准商铺的设计延续花砖、铁艺灯元素，又以天蓝色、鲜黄色等亮丽的色彩加以碰撞，让生活购物区散发着浓厚的生活气息、邻里氛围，由此释放出社区的无限活力。

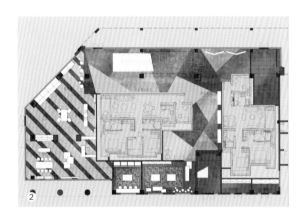

Composite space of multi-functions and community-based sales center

Besides visually appealing, a community-based sales center is more in need of practicality. Therefore, the designer reserved spaces for multiple life and business possibilities, where a composite space would be quite helpful.

Regular stores continue to use tiles and iron art lamps with a touch of bright colors like sky blue and yellow, making the shopping area more down-to-earth, vigorous and amicable.

❶ ❺ 烘焙区弥漫着甜蜜的味道。
   The baking area.
❷ 商店平面图。
   Layout of the shop.
❸ 商店立面效果图。
   Elevations of the shop.
❹ 花店里繁花似锦。
   The flower shop.
❻ 休闲区在阳光中显得浪漫。
   Leisure time.

在海的近处，城市的核心，一处形似鼻舳的前卫建筑似乎在强调着它不一样的建筑功能。深圳海上世界私人沙龙会所拥有丰富的空间资源和景观视野，收藏着来自世界各地的文化藏品，展示着收藏家非凡的眼光及品位。设计利用高空优势和光线布局，结合窗外粼粼的水色景观，回归材质的原始状态，简洁有力地赋予收藏空间有序的展示阵列，最大程度地烘托藏品的价值和独特美态。

With rich space resources and landscape views, Shenzhen Sea World Private Salon Club keeps cultural collections from all over the world, revealing the collector's extraordinary vision and taste. Leveraging the building's advantages of high altitude and lighting conditions, combined with the aqua landscape outside the window, the designer maintains the orderly array in the collection room, maximizing the value and unique beauty of the collections.

① ④ 众多的私人藏品按照年代、艺术方式进行陈列。访客可随着建筑的"鼻舳"体态，拾阶而上。
Numerous private collections are displayed in a chronological and artistic manner. Visitors can climb the stairs following the bulbous bow of the building.

② ③ 众多的藏品让人目不暇接。
Many collections make a dazzling.

⑤ 建筑俯视图。
Architectural elevation.

④

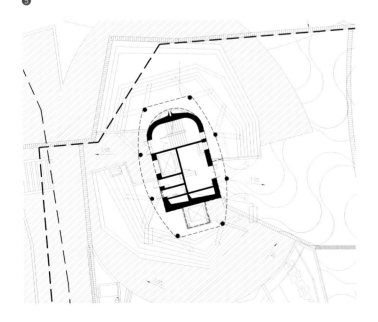

⑤

由于藏品众多且形态不一，室内设计方面优先考虑陈列方案，务求让每件珍贵的藏品都能找到自己的安身之所。一层中央展示区的墙体全部用作展示，用以收纳绘画作品、工艺藏品、标本、书籍等。小件石雕则组合成空间巨大且别致的吊饰群，增添了空间的视觉趣味。腾出的地面空间留予访客游走、观赏和沉思。

With a huge amount of collections in different forms stored here, the interior design prioritizes the layout of the room so that every precious collection can find its own place. The walls of the central exhibition area on the first floor are used for displaying paintings, craft collections, specimens, books, etc. The small stone carvings are combined into a huge and chic drop ornament group, adding visual interest to the space. And the vacated ground is reserved for visitors to walk, watch and contemplate.

❶❷❻❼ 建筑突出的前端如同航船的甲板，临近海上世界著名的明华轮，抬眼即可将水色夜景尽收眼底。
Close to the world-famous Anceveller, the building with its protruding front edge resembling the ship deck enables one to just look up and enjoy beautiful sea views and night views.

❸❹❺ 临窗处的圆弧、电梯井附近以及悬空设计的三个独立阅读空间均设有卡座供阅读和休憩，空间的功能和景观优势得到综合和提升。在沉思中，人们与思想和艺术同航远方。
Booths equipped for reading and resting in the three independent reading spaces, namely the arc-shaped area near the window, the area near the elevator shaft and the suspended area....all these arrangements strengthen the functions and landscape advantages of the building.

位于海上世界负层导购区艺术品商店根据藏品的年份依次陈列，由此得到一卷实体的文化编年史。特别设置的吧台和卡座意识导购区的一景，浓郁的工业风情让人沉浸其中。

The art store, located in the shopping area on the basement floor, displays its collections in line with the according years, thereby creating a cultural chronicle in real terms. The bar and the booth area are also sceneries in the shopping area, delivering an attractive industrial style.

❶❹❺ 在展示厅中，整体简约、气氛复古的设计手法贴合收藏品的调性，跳跃的色彩的运用则缓和了较为沉静的氛围。
The design approach of overall simplicity and vintage atmosphere fits into the tonality of collections, while the solemn atmosphere is eased using the vibrant colors.

❷ 展示厅入口。
The entrance of the exhibition hall.

❸ 展示厅平面图。
Layout of the hall.

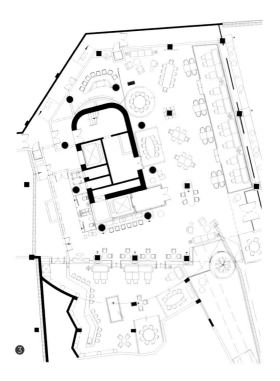

④

⑤

启迪协信杭州科技城展示中心
EXHIBITON CENTER OF TUSINCERE SCIENCE CITY, HANGZHOU

关键词：未来感、城市展厅、智能化
Keywords: futurism, urban exhibition hall, intelligence

总建筑面积 / 800,000 平方米
委托范围 / 硬装设计
委托面积 / 1,400 平方米

流动之美

曲面的流线墙体，立体的艺术照明，互动的的流动光影，智能系统的加入融合成一个多功能的未来空间。艺术与科技的融合，让空间更富立体感与未来感，内部的线条阵列及变异塑造了别样的光影之美，让人沉醉其中。

Flowing beauty

The cambered and streamlined walls, three-dimensional art lighting, interactive flows of light and shadows, and the intelligent system are integrated into a multi-functional future space. The fusion of art and technology makes the space more three-dimensional and futuristic while the internal array of lines helps to create a different kind of light and shadows, making people indulge in it.

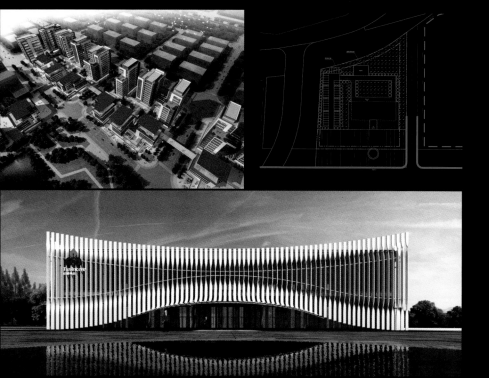

白色盘旋，点亮空间精气神

首层统一以白色为主要的空间色调，流动的光影随着曲面的墙体在空间延伸，室内流动的曲线设计与地面的藏光地灯引领空间的动线。沙盘区中空吊灯似紫荆花散落的花瓣，集聚汇之势盘旋上升。水吧区则以科幻感家具为主，将等候区布局在光的消散之处，为来访者提供休闲之所。

White lights the space

The first floor applies white as the main tone with the flowing light and shadows extending on the cambered wall and flowing curves and hidden lights on the ground leading the overall flow of the space. The chandelier in the sand table area is like petals scattered from Chinese redbud, rising in a spiral manner. The bar area features in sci-fi furniture, and the waiting area is laid out in the place where light disappears, providing visitors with a relaxing place.

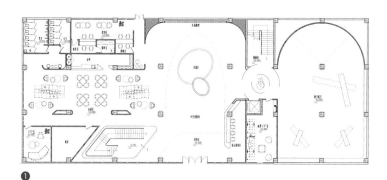

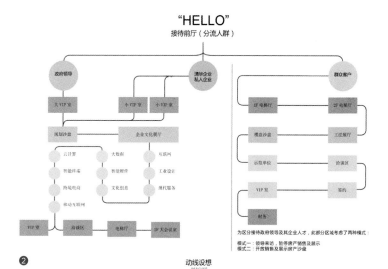

① 展示中心首层平面图。
Plan of the first floor.
② 动线规划。
The route planning.
③④⑤ 宽阔的接待区。
The spacious reception area.
⑥⑦ 在设计之初便提前考虑到智能系统的应用，为空间预留更多可能性。
The application of the intelligent system is considered at the beginning of design for reserving possibilities to the space.

智能互动,引领风尚

未来的城市需要变得更加智能化以及高效,其环境能够应对未来的各种变化,因此设计师结合企业文化和建筑语言融入到空间当中。"未来展厅"从入口的设置就加入了人像的互动窗口,走入室内如同置身于未来之城。智能系统的运用,互动展示模式让人耳目一新,历史文化中心,科幻式的灯具装饰和大型的互动荧幕,展示着企业对未来的规划与展望。

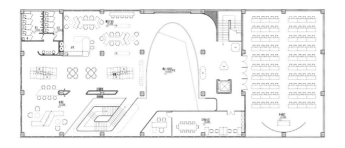

Smart interaction, leading the fashion

Cities in the future should be more intelligent and efficient with their environment able to cope with future changes, so designers integrate corporate culture and architectural language into the space. The "Future Exhibition hall" adopts the interactive window of portraits at the entrance, and visitors would feel like being in the city of the future. The application of intelligent systems and interactive display mode is refreshing. What's more, historical and cultural centers, sci-fi lighting and large interactive screens all display the company's future planning and outlook.

❶ 二层休憩接待区。
The reception area of the second floor.
❷ 二层平面图。
Plan of the second floor.
❸❹❺ 智能展览区。
Exhibition area with intellectual technology.

上海协信集团企业会所
SINCERE CORPORATE CLUB, SHANGHAI

关键词：海派文化、西式风华、中式线条
Keywords: Shanghai-style culture, western style, Chinese style

建筑设计 / Aedas凯达环球事务所
景观设计 / 棕榈设计
建筑面积 / 225,622 平方米
委托范围 / 硬装及软装设计
委托面积 / 2,332.7 平方米

当我们谈及上海，这个集东西情韵于一身的地方，古典、雅致同时又有国际大都市的现代与时尚，具有开放而自成一体的独立风格。上海特有的海派文化，植根于江南的吴越文化上，融汇中国其它地域文化的精华，吸纳西方的文化因素，吸纳百川，勇于创新。

电梯缓升至9层，门打开的瞬间，柔和灯光下有着简约的金色线条，白色的大理石地板，现代感十足。接待厅内旋转的楼梯，被分割成一个个小正方形的玻璃天花，线条感十足。

Shanghai, vintage and elegant as an old city while modern and fashionable as an international metropolis, has formed an open and independent style. Shanghai's unique Shanghai-style culture, rooted in the Wuyue culture of the south of the Yangtze River and blended with the essence of other regional cultures in China and western culture, is inclusive and creative.

With the elevator rising to the 9th floor, the moment when the door opens, one can see simple gold lines under the soft lighting, white marble floor and spiral staircases in the reception hall divided into small square glass ceilings, delivering a modern atmosphere.

① Golf practice field 高尔夫练习场	⑨ Fold the aqua green wall 叠水绿墙
② Stage 舞台	⑩ The tree and stone 花树景石
③ The counter 备餐台	⑪ Six people teahouse 6人茶室
④ Dinng room 聚餐场地（舞池）	⑫ Six people teahouse 6人茶室
⑤ Dinng room 10人餐（茶）室	⑬ Tree on the scene 对景花树
⑥ The pond 水池	⑭ Cascading water features 跌水水景
⑦ Water plantform 水上表演台	⑮ Stone steps 景石台阶
⑧ The garden entrance 花园入口	⑯ Viewing platform 观景平台

❶❷❸ 进行顶层空间设计时将园林景观纳入设计范畴，对外空间与内空间进行整合性设计。
The landscape is incorporated into the design range during the design of the top-layer space. The external space and the inner space are designed in integration.

❹❺❻❼ 奢华古典的配饰令空间气势非凡。
Luxurious and classic furnishings.

在会所的不同空间内，随处可见的是充满吴越经典的装饰设计，江南烟雨，山水林立。茶室内红棕色的木桌，瓷器质感的圆凳，中式的装潢配饰，更加映衬了空间的文化沉淀。头顶天花设计以及简约的中式家具，摈弃了古风装饰的复杂，采用现代简单干练的线条。多样的主题宴会厅，多个文化古都的精髓与西方文化元素的结合。轻奢新中式家具与意式家具的碰撞，诠释让人出乎意料的品质。因此我们在会所整体的设计风格上，除了有着江南风情的吴越文化元素还有近现代西方的简约和美式Art Deco的风格，现代与经典的融合，在细节中感受奢华。

In different rooms of the club, decorations of Wuyue cultural elements can be seen here and there. The red-brown wooden tables in the tea room, the porcelain-style stools, and the Chinese-style decorations reflect the cultural deposits of the space. The ceiling and Chinese furniture apply modern simple and neat style instead of complicated ancient decorations. The banquet halls with various themes are a combination of the cultural essence of multiple ancient capitals and the western culture. As the integration of luxurious new Chinese furniture and Italian furniture interprets unexpected quality, thus we connect Wuyue culture style with modern American Art Deco style, the fusion of modern and classic, allowing visitors to enjoy the luxurious details.

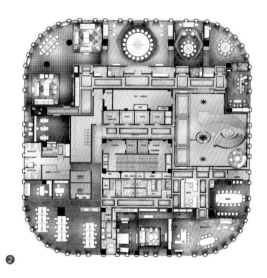

现代中式相比欧式的繁华，美式的利落，有一种繁简兼具的优点，如太极之道，恰到好处的利用方能发挥设计的最大视觉呈现。本案以现代简约为主基调，根据空间功能适度地将风格元素贯穿其中，恰到好处的色彩运用、配饰选择、硬装把控令不同空间的设计多元融合。

Compared with prosperous European style and neat American style, modern Chinese style is a combination of the two. For example, the proper use of Tai Chi can maximize the visual effects of designs. This case uses modern minimalism as the main tone, decorated with style elements in accordance with the space functions, and the right color application, proper decorations and hard-fit control integrate the designs of different spaces.

① 设计气质稳重的电梯间。
The elevator room.
② 会所平面图。
Plan of the club.
③④ 茶艺室里飘逸着中式美学。
A tea room full of Chinese aesthetics.
⑤⑥⑦ 不同主题的豪华餐厅。
Different themes of the luxury restaurant.

深圳贝骊洛生活美学馆
BEST & REAL LIVING ZONE, SHENZHEN

关键词：生活美学、跨界整合、工业风
Keywords: aesthetics of life, cross-industry collaboration, industrial style

项目面积 / 2,700 平方米
设计范围 / 硬装及软装设计

浮躁的商业社会，正需要我们用心去享受生活的美好，深圳贝骊洛美学生活馆秉承着"Best&Real"的原创理念，以多元精致的设计风格，融合性空间布局专注于每一个当下。综合咖啡轻食、服饰配饰、图书文创、家具纺织、意式西厨等多种业态，打造乐享慢活态度的美学体验空间，呈现文化艺术与创意灵感的多重精彩。目光尽处，觉美之旅。衔接生活，一馆蔽之。

Shenzhen Beililuo Lifestyle Store honors its original concept of "Best & Real" and focuses on every present moment with diverse and exquisite designs and integrated layouts. It covers café and light meals, clothing and accessories, books and creative goods, furniture and spinning, Italian food, and etc. It aims to offer aesthetic experience of "Slow Movement", presenting the wonderfulness of culture, art, and creativity.

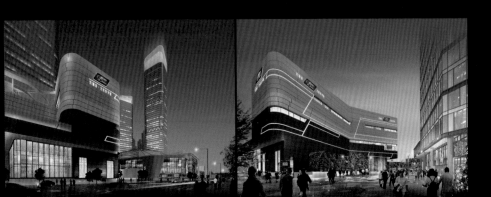

精品生活大区的回环连廊串联起不同类别的展示空间，让人目不暇接。浅金色与浅灰色营造出轻盈、时尚且优雅的空间氛围，以巨大的灯饰为纽带，带动中空两层的交互呼应。

The winding corridor of the Exquisite Life Area connects different display zones. The light gold and grey tone create a relaxing, fashionable, and elegant atmosphere, coupled with a huge chandelier promoting the interaction between two floors.

❶❷❸❹❺ 利用建筑的跨层优势，按照消费者习惯动线打造错层式的销售展示空间。
Taking the cross-layer advantages of the building, a split-level sales display space is created in accordance with the motion line of consumer habits.

SNP DESIGN RECORD VOL.2 295

意式饮食，有滋有味

意大利设计美学一直坚持艺术与实用哲学的高度结合，这一点在贝乐厨意式餐厅得到了充分的体现。开放式的厨房和操作台为主厨团队提供一个最广阔的舞台。设计师在这里着重强调了经典的黑白冷暖对比的同时，又加入装饰性的折线和色块等元素，带来别样的进餐体验，让顾客享受到地道正宗的意式美食。

Tasty Italian cuisine

Italian designs value the combination of art and practical philosophy, which is vividly embodied in Bere Italian Restaurant. An open kitchen and operation desk provide a wide stage for the culinary team. The designer highlights the contrast of "black and white" and "cold and warm tones", coupled with decorative broken lines and color blocks, allowing customers to enjoy special and authentic Italian food.

❶❹ 餐厅就餐区及VIP室效果图。
Rendering of the dining area and VIP room.
❷❺ 就餐区实景。
Pictures of the dining area.
❸ 二层空间平面图。
Plan of the second floor.

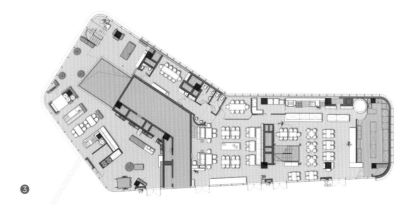

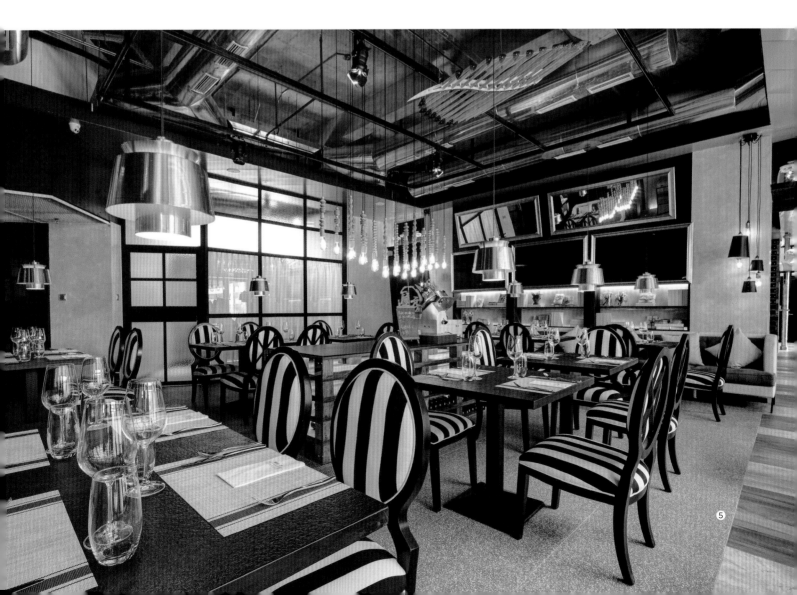

❶❻ 商店橱窗外观效果图。
　　Rendering of the display window.
❷ 首层平面图。
　　Plan of the first floor.
❸❹❺ 图书区、文创区及咖啡区。
　　The book, cultural products and coffee area.

轻阅读，慢品味

醇香的咖啡，图书墨香以及悠扬的乐声渲染着优雅气场。简练的工业风格充满随性和自由，各种元素在这里都能找到自己的故事和节奏。黑色铁艺的沉稳与亮的金属，不露痕迹地划分了展示区和休憩区，翻一页书，饮一口咖啡，品味生活的写意，时间也在此伫足。

Light reading and relaxing life

The aroma of coffee and books as well as the sound of music deliver an air of elegance. Simple industrial designs are filled with ease and freedom. The composed black iron art and bright metals naturally separate the display area and recreational area where one can enjoy reading, coffee, and a relaxing life.

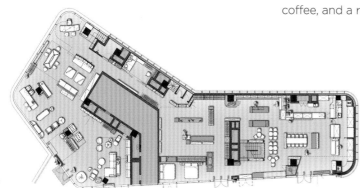

人与居的装饰美学

家具区呈弧形布局,用灰色调及裸露的清水漆墙体诉说着自然质朴之美。进口家具本身的排布组合形成一个又一个的"面块"让空间分区明确又无明显隔断,方便导赏和走动;延至服装区,高低错落的置物层架采用锐利明确的直线条以及大面积的玻璃或亚克力。俗话说,人靠衣装。空间装饰材料的严谨与衣服的飘逸形成了对比,巧妙地营造出"快时尚"的质感,让人置身其中不经意间亦产生共鸣。

Human-friendly and living-friendly decorative aesthetics

Furniture area is designed in an arc shape, coupled with a grey tone and natural non-oil painted wall. The layout of imported furniture naturally forms blocks—functionally separated yet interdependent and unobstructed. Layers of shelves in the clothing area apply sharp lines and large-scale glass or acrylic. The contrast between the rigor of decorative material and the softness of clothes creates a sense of "fast fashion" and resonating feelings.

❶❸ 进口家具展示区。
The import furniture exhibition area.
❷ 服饰区。
The clothing area.

别墅
VILLA

别墅是品质生活的重要载体。基于空间的尺度和独一无二的资源，为各种生活情境提供无与伦比的舒适度，承载成功人士对生活情怀的远见。
Villa is an important carrier of quality life. With the spaciousness and unique resources, it provides unparalleled comfort for various life settings and holds the vision of successful people for future life.

重庆保利江上明珠依山别墅
北京保利首开天誉别墅
香江海滨私人别墅
Villa Of Poly Real Estate, Chongqing
Villa Of La Vita E Bella, Bejing
Private Villa On Pearl River

重庆保利江上明珠依山别墅
VILLA OF POLY REAL ESTATE, CHONGQING

关键词：生活品质、尺度感、优雅
Keywords: quality life, propriety, elegance

景观设计 / 重庆曲线景观
建筑面积 / 74,360 平方米
委托范围 / 硬装及软装设计
项目面积 / 480 平方米

奢华的居所设计中，除了选材，功能的合理分布及恰到好处的装饰亦十分重要。坐落于重庆这个自然与都市交融一体的经济新城，保利江上明珠别墅把以上三者合理组合，将空间层层剥开，重现品质生活之美。

For designs of luxurious residence, besides material, a reasonable layout of functions and appropriate decorations are also of crucial importance. Located in Chongqing—a new economically emerging city enjoying the beauty of nature and urban development, villas of Poly on the River Pearl cleverly combine the above three and present the beauty of quality life.

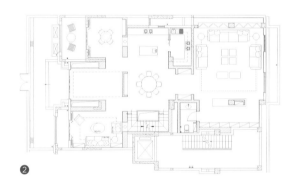

无论是主人归家或朋友来访，别墅玄关作为入户所见的第一重展示空间都显得十分重要。项目前厅宽阔端正，又有适度的进深，我们利用镜面效果将其设计成一个小而美的艺术展示区以流露主人的独特品位。玄关中央起到类似中国传统宅邸"影壁"的作用，正面悬挂艺术画作，背面亦可以起到收纳架的功能，强调突出了"门户"的尊贵感。

- ① 别墅玄关。
  The entrance area.
- ② 首层平面图。
  Plan of the first floor.
- ④⑤ 精致的艺术品。
  A variety of delicate artworks.
- ③⑥ 首层客厅。
  The living room on the first floor.
- ⑦ 餐厅景致。
  The dining room.

Be it returning home or visiting, the hallway of a villa is the first and foremost impression of the person who sets foot into the house. The lobby in this case is spacious with moderate depth. We apply mirror effects and turn it into a small but exquisite exhibition area to display the master's taste. The center of the hallway, similar to the "screen wall" of traditional Chinese residence, hangs art painting on the front and serves as storage racks on the back, highlighting the household's prestige.

负层是别墅的主要娱乐空间,水吧、桌球娱乐、健身房、会议室一应俱全。颇具设计感的旋转门处理为空间添注了新鲜感和时尚感。蓝色及紫色的灯光搭配契合空间特性,狂野迷人。

The basement floor serves as recreational area with bar, billiards, fitness room, and conference room all in one. The revolving door adds an air of freshness and fashion to the space, while the blue and purple lighting fits right into the space-- wild and appealing.

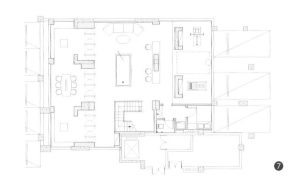

❶ 负层楼梯间。
　The stairs of basement.
❷❸❹❺ 会议室景致。
　The meeting room.
❻ 娱乐室景致。
　The recreation room.
❼ 负层平面图。
　Plan of the basement.

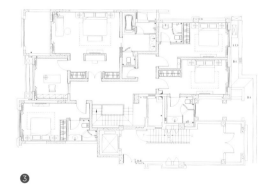

❶❹ 豪华的主卧。
The luxury master room.
❷ 优雅精致的陈设。
It's full of elegant and delicate furnishings.
❸ 二层平面图。
Plan of the second floor.
❺ 次卧景致。
One of bedrooms.
❻ 书房景致。
The study.

二层主要为卧室休憩区。四个套房均拥有绝佳的赏景角度，尽显别墅最大的优势——每个家庭成员都能拥有自己的私密天地而不互相打扰。

The second floor is for bedrooms. Four suites with fantastic views highlights the advantages of a villa—every family member gets his/her own secret space without disturbing each other.

北京保利首开天誉别墅
VILLA OF LA VITA E BELLA, BEJING

关键词：有闲阶级、混搭风、工业美学
Keywords: leisure class, mixed style, industrial aesthetics

建筑设计 / HZS汇张思
建筑面积 / 123,579 平方米
委托范围 / 硬装和软装设计
委托面积 / 650 平方米

保利首开天誉位于朝阳区东坝区域，位居CBD商圈，临空经济带，以"为城市高端客户打造城市高端居住区"为理念进行当代演绎，为拥有独特时尚品位的精英人士打造相得益彰的居住空间。

我们实地考察北京城区内多个别墅区，发现其中大多数的装饰风格较为华丽。因此我们另辟蹊径，决定在整体布局或装饰陈设上均创新求变，务求令这个纯正的法式别墅建筑拥有自己独一无二的气质和风格。

Poly and BCDH Villa is located in the Dongba area of Chaoyang District, Beijing. At the heart of CBD and adjacent to economic zone, it aims to create a unique and matching residence for elites with fashionable taste.

We visited a number of villas in the urban area of Beijing, finding that most of them adopt a more ornate style. So we have taken a different path and decided to innovate and change the overall layout or decoration, so that the pure French villa has its own unique temperament and style.

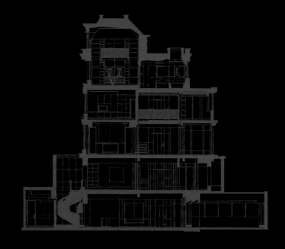

地下开放的华丽空间

从生活动线的角度看,住户与拜访者一般从地下车库进入别墅,地下空间的第一印象尤为重要,因此对空间与光的运用则成为我们整个地下空间的思考重点。

为了减少地下空间带来的密闭感,我们打通负一层、负二层的楼层隔断,在原建筑的三个采光井(前院采光井、采光井及后院采光井)的基础上作无间处理之外,又把地下空间原有的窗框全部剔除,最大限度把地下空间整合到一体,方便自然光线充盈于此,保证室内空气对流畅通无碍。

地下层空间感与功能布置已得到满足,我们还要为空间塑造出一道合适有闲阶级趣味的文化景观。我们结合了电影布景,将工业美学的元素与建筑中古典美学元素有机结合,相互交融,让整个空间设计给人一种视觉的冲击感。

Gorgeous and open underground

Residents and visitors generally enter the villa through the underground garage, making the first impression of the underground particularly important. Thus, the layout of space and light are the focus of the underground design.

To soften the sense of airtightness brought by the underground, the designer connects -1F and -2F, conducts non-barrier treatment based on the three light wells (forecourt light well, light well and backyard light well) in the original building and removes the original window frames, integrating the underground space to the fullest. In this way, it helps to let the natural light come in and ensure the unobstructed air flow.

We also create a cultural landscape suitable for the leisure class as the spaciousness and functional layout of the underground have been ensured. Coupled with the filmset, we fuse the industrial aesthetics with the classical aesthetics of architecture, making a visual impact.

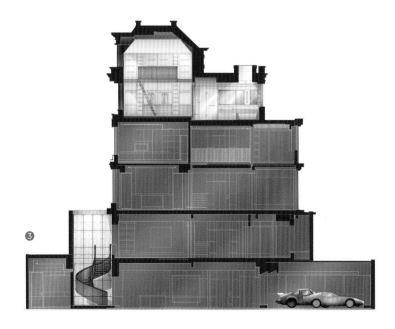

❶❷ 通往负层的楼梯间。
The stairs leading to basement.
❸ 建筑剖面图。
Architectural section.
❹❺❻❼ 地下室的休闲美学。
It's full of leisure atmosphere in the basement.
❽ 空间手绘图。
Space sketches.

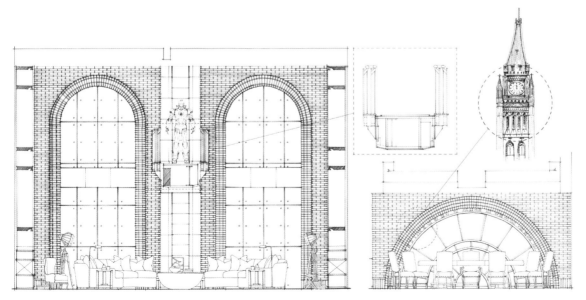

更贴合家庭观念的首层客厅

首层是家庭生活及接待的主要空间，设计师以天然材料、质感色彩和利落的线条来塑造，简约中有艺术品位、结构中深藏形式美。带有工业复古气息的陈设令空间顿生沧桑感，表达了主人对于家居世界中不同价值观的探索。

为了满足不同场景的需求，设计师利用空间宽敞的优势，客厅使用了多人聚拢和分区交流的布局。客厅并未有过多地强调造型与装饰，反而更多地留给客户去感受空间的本身。另外，通过朴实的色调、温润的质感及家具的品质带给客户舒适的氛围，体现出对客户心理的微妙考虑。

❶❸❹ 宽敞舒适的客厅。
　　The wide and comfort living room.
❷ 客厅手绘图。
　　Sketch of the living room.
❺ 细腻精致的陈设品。
　　Details of delicate furnishings.
❻ 首层平面图。
　　Plan of the first floor.

A more down-to-earth first floor living room

The first floor is the main space for family life and reception. The designer uses natural materials, textured colors and neat lines to create artistic simplicity and beautiful structure. The furnishings with industrial vintage style give the space a sense of vicissitudes and show the owner's exploration of different values in the home.

For the living room, the designer utilizes the spaciousness of the room and arranges it with a layout that meets the needs of different occasions like a gathering or separate communication. It does not emphasize the style or decoration too much, instead allows the clients to feel the space itself. The simple color, mild texture and quality furniture create a comfortable atmosphere for the residents.

❶ 中空设计的餐厅格外别致。
The dining room with open hollow design.
❷ 首层空间充满了艺术气息。
It's full of artistic atmosphere on the first floor.
❸ 设计师对精英生活模式进行深入了解，设计出或独立或开放的聚会客厅，以打造适合不同人数的交流空间。
Independent or open get-together living rooms are designed on the basis of in-depth understanding of elite living modes, so as to create a communication space for different people.

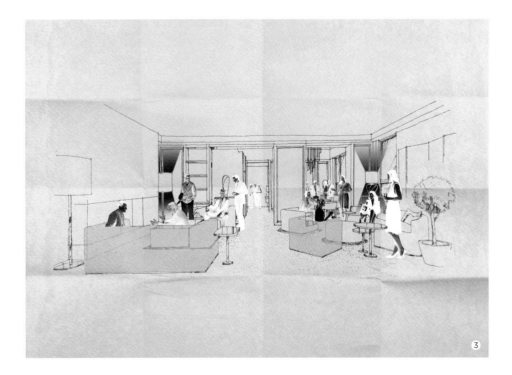

简约复古的素雅空间

有别于一层的聚拢热闹,二层显得更为素雅温馨。在走廊处可见上下两层相连的中空区域,充分展示了建筑的层高,在此可与一层空间无碍互动。长者房静卧茶案一座,一点天然,一点复古,一点简约,皆成气氛。

Simple and vintage space

Different from the lively first floor, the second floor is more elegant and warm. In the corridor, the hollow space between the two floors demonstrates the height of the building, creating unobstructed interaction between the two floors. A tea table lies in the senior's room, creating a natural, vintage and simple atmosphere.

❶ 简约舒适的次卧景致。
A clean and comfort landscape in one of bedrooms.
❷ 二层平面图。
Plan of the second floor.
❸❻❼ 二层楼梯间。
The stairs.
❹❺ 简洁陈设烘托着卧室宁静的氛围。
Simple designs tend to foiling the quiet atmosphere in the bedrooms.
❽ 二层洗手间。
The bathroom.

何谓拥有顶级尺度的尊贵主卧？

主卧空间对女主人而言，衣帽间和浴室是是精美的收藏馆和难得的放松之所。原设计中阁楼光线不足，给人一种压抑感；侧旁的楼梯稍显狭窄，浪费了部分空间。因此我们借鉴电影《了不起的盖茨比》，打造一个京城地带独一无二的复式回廊华服收藏馆。这个改动巧妙地整合空间，让原有的衣帽间得到更好的舒展，营造别墅主卧豪华的顶级尺度感。

Luxurious main bedroom

For the mistress, the cloakroom and bathroom in the main bedroom are for exquisite collections and relaxing respectively. In the original design, the attic was in short of light, creating a sense of oppression; the adjacent stairs were a little narrow, a waste of space. Thus, inspired by the film The Great Gatsby, we created a unique double-corridor collection room in the capital city. This adjustment integrates the space and allows the original cloakroom to be extended, creating a luxurious and spacious master bedroom in the villa.

❶ 顶层主卧具有绝佳的舒适度和私密性。
The master bedroom has excellent comfort and privacy.
❷ 衣帽间也可以拥有大尺度感。
A wide cloakroom.
❸❹❺❻ 卫生间同样综合了建筑的原有特点，特意在浴缸之上设置了天窗，让内外风景得以互动。沐浴时洗去疲惫，还能抬头瞥见星空，岂不美哉。
The bathroom maintains the original features of the building but adds a skylight above the bathtub, realizing the interaction between interior and exterior views.
❼ 顶层剖面图。
Section of the top floor.

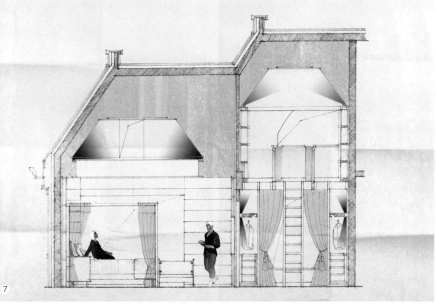

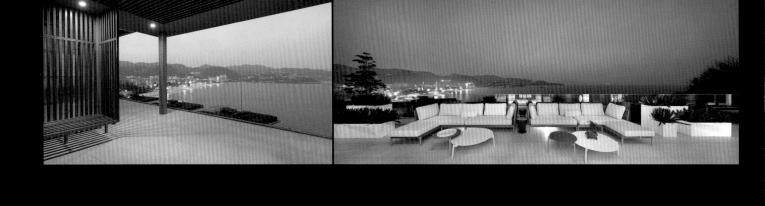

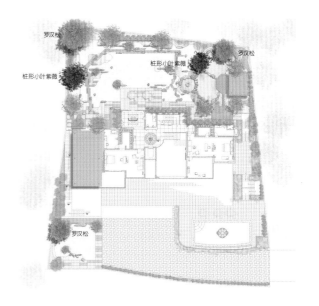
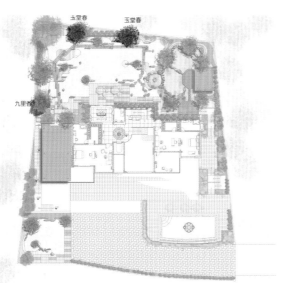

亭亭之景，天海一色

海浪澎湃，气象大开；园林扶疏，错落成趣。极具尺度感庭院如一副待上色的画卷，SNP首次执起画笔，细细描摹一个生趣盎然的游园。一株古松立于门庭，如谦逊有礼的管家迎接主人。延至后院，富有禅意的枯山水韵味无穷。所栽种的松柏花叶形态各异，临风飒飒，蕴藏着无穷的生命力。绿意的闹，白砾的静形成了有趣鲜明的对比。茶亭掩映其中，沉静安然。面朝大海，春暖花开，于此处听海声，读风月，把生活过成了诗。

❶ 园林设计的推敲过程。
The evolution process of landscape design.

❷❸❺❻ 优秀的建筑条件带给设计师更多想象空间。设计师在建筑主体的侧面设计出一个独特的休憩庭廊。该庭廊前见大海，后倚园林，左接室内前厅。建筑、景观、室内借一个半开放空间收束到一条线上，十分连贯且一气呵成。
More spaces for imagination can bring to the designer with excellent building conditions. A unique lounge is divided from the side of the main part of the building. In the corridor, you can see the sea in the front, the garden on the rear and the interior hall way on the left. The building, the landscape, and the interior are closed to a line by a semi-open space, which is coherent without stopping.

❹ 别墅园林景观与大海相依。
Landscape and seascape.

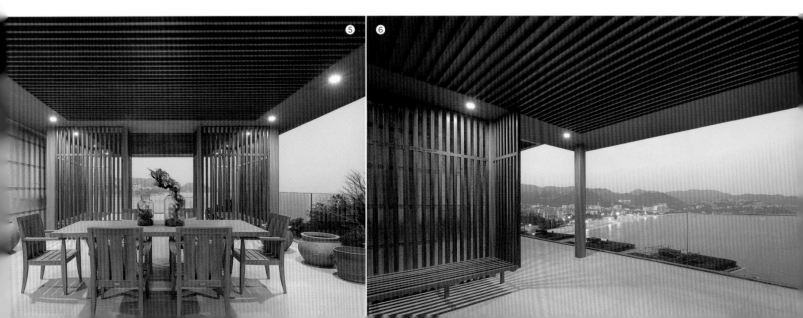

Views from the pavilion: the sky and the sea melt into one

The spacious courtyard is like an unpainted scroll where SNP picks up the brush for the first time and paints a lively garden. An old pine stands at the courtyard like a modest and polite butler ready to welcome guests. The backyard is a dry landscape garden, delivering an air of "Zen". Pines, cypresses, flowers, leaves, etc. are in different shapes and full of vitality. The vigor of green and the stillness of white gravel form an interesting contrast, where a serene pavilion hides within. Residents can sit here and listen to the sea, living an aesthetic life.

❶ 园林平面图。
Plan of the garden.
❷❸ 枯山水的景观设计与建筑本身的石材完美结合，实现建筑的多角度观景需求。
The landscape design of Japanese rock garden is perfectly combined with the building stones, realizing the multi-angle viewing demand of the building.

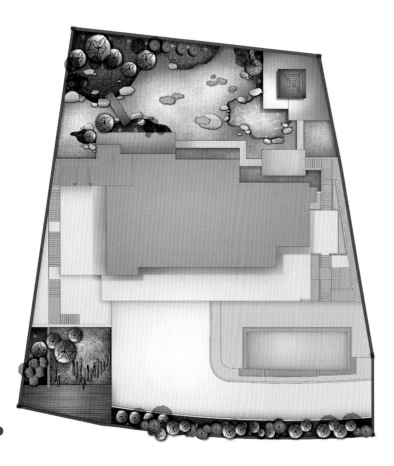

面朝一线海景，这所家宅的每一层配有相应的活动空间。室内室外不断互动，自由连接，也增加视觉上尺度感和实用灵活性。折叠式落地玻璃门完全敞开时，处于负一层的室内健身房自然而然地连通了户外平台。健身不再囿于"一室"或"一房"，无论挥洒汗水还是碧波畅游，更广阔的平台无疑都放大了运动的乐趣。接待客人时，这处空间也大有精彩：既可以相互碰杯，静看一池清水倒映着蓝天白云；也可以举办一场盛大泳池+BBQ派对，享受夏日欢乐。

With full sea views, every storey of the villa has corresponding functions. The continuous indoor and outdoor interaction improves the spaciousness visually and flexibility practically. When the folding French window fully opens, the fitness room at the basement naturally connects with the outdoor platform. In this way, fitness is no longer confined within "one room" and a wider stage undoubtedly adds more fun to working out. And this space reserves more surprises when receiving guests: people can either just gather here and enjoy the beautiful views or hold a big pool+ BBQ party and enjoy the fun of summer.

❶❷ 透过落地玻璃窗的开合，户内户外无缝连接。
Outdoor and indoor are connected by the glass windows.
❸❹❺ 水吧及游泳池景致。
The bar and swimming pool.
❻ 负层平面图。
Plan of the basement.
❼ 首层平面图。
Plan of the first floor.

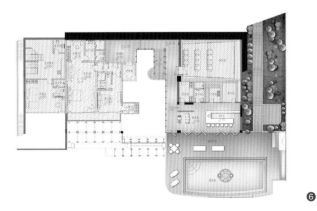

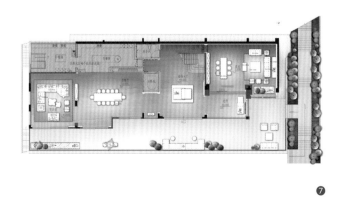

❶❷ 首层玄关效果图及实景。
Rendering and picture of the hall way on the first floor.
❸❹ 宽阔的餐厅景致。
The wide dining room.
❺ 雪茄室景致。
The cigar room.

一层也是社交聚会的好地方。沿着超大尺度的面宽，设计师在首层分别安排了视线互通的雪茄室、餐厅、大厅和茶室。新亚洲的装饰风格令这几个空间保持统一时也拥有自己特别的个性。雪茄室中的意大利 MINOTTI 沙发令空间显得优雅舒适。餐厅则预留了生活模式和宴会模式。主人可以为客人"定制"不同的社交场景。别墅南向的全落地玻璃幕墙设计肯定能为客人呈现最完美的海景和自然天光。

The first floor is a good place for social gathering. Leveraging the width of the floor, the designer sets up a cigar room, a dining room, a living room and a tea room with mutual transparent views on the first floor. New Asian decorative style allows these rooms to be consistent with each other while reserving their own individuality. The Italian MINOTTI sofa makes the cigar room more elegant and cozy. The dining room reserves for parties or other occasions where the master can tailor-make different social settings for guests. The south-facing floor-to-ceiling glass curtain wall at the villa can definitely show the best sea views and landscapes for guests.

❶❸ 办公室的设计沉稳且富有格调。
The decoration style of the office is steady.
❷ 首层客厅。
The living room on the first floor.
❹ 二层平面图。
Plan of the second floor.
❺ 顶层平面图。
Plan of the top floor.

现代豪宅渐渐出现了对办公室、会议室等商务空间的需求。我们顺应趋势，整合项目本身的空间特点，预留了类似的功能区间以满足主人所需。

Modern luxurious residence has been gradually in need of business space like office and conference room. Catering to this need, based on the space characteristics of the project itself, we reserved similar functional rooms for clients.

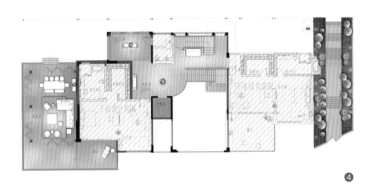

大海的摇篮曲

二层和三层是重要的起居空间。卧室延续着新亚洲的装饰风格,用色更为沉静。每当夜幕降临,大海的摇篮曲从远处绵绵传来,浪漫的星光入梦,甜美安静的夜晚总令人安眠。院宅府邸既是生活审美的载体,也深刻影响着家族传承的脉络。和风细浪是大海的歌咏,而诗意则是专属人类的咏叹调。一座府邸成了诗,而生活在里面的人,也变成了诗的一部分。

❶❷❸ 聆听大海的安眠曲,安然入梦。
When the night comes, natural lullaby from the sea along with the starry sky accompanies one to sleep.

❹ 舒适柔软的灯光设计。
The tranquility and thoughtfulness for resting with a comfortable lighting atmosphere.

❺ 浴室里也能观赏最佳的海景。
Overlooking the sea in the bathroom.

Cradlesong from the sea

The second and third floor are important living space. Continuing to use new Asian decorative style, the color tone would be more composed. When the night comes, natural cradlesong from the sea along with the starry sky accompanies one to sleep. A residence is a carrier of life aesthetics as well as family inheritance. Light winds and waves are hymns to the sea, so is poetry to human beings. A residence turns into a poem, which makes the residents part of the poem.

现代主义
MODERNISM

我们深知"少即是多"的现代主义美学精髓，摒弃过度的修饰和多余的形式感，着重强调空间功能、综合能力和实用性，以利落明快的手法为项目的气质量身订造富有个性的设计细节。

We are fully aware of the essence of modernism aesthetics of "less is more", namely, abandoning excessive ornaments and superfluous sense of form, emphasizing the comprehensive ability and practicality of spatial functions, and tailor-making a unique design of the project with a clear and bright approach.

重庆保利江上明珠公寓
合肥保利海上明悦销售中心
莆田保利拉菲公馆销售中心
北京和锦薇棠销售中心
清远保利春晓接待中心
Loft Of Poly Real Estate, Chongqing
Sales Center of Poly Sea Bright, Hefei
Sales Center Of Poly Lafite Mansion, Putian
Sales Center of Poly Belle Ville, Beijing
Reception Center of Poly Spring, Qingyuan

重庆保利江上明珠公寓
LOFT OF POLY REAL ESTATE, CHONGQING

关键词：简约、艺术感、分寸感
Keywords: simplicity, art, sense of propriety

景观设计 / 重庆曲线景观
建筑面积 / 74,360 平方米
委托范围 / 硬装及软装设计
委托面积 / 160 平方米

Lofter相对小体量、灵活的特点，契合当下的现代化生活模式和市场结构。巧妙利用高差，整合较为零碎的空间以布局空间。对具备复合功能的空间要避免过多的标签，而应该采用简约化手法，令其因地制宜，保证居家的舒适感和尺度感之余仍不失个性和趣味，这都是室内空间设计实现以小见大的门道所在。

简约并不意味着容不下一丝额外的装饰。我们依据主人的性情恰到好处地为空间添置内心的向往和生活的痕迹，是室内设计的精髓所在。重庆保利江上明珠公寓无论在空间处理还是装饰布置上都保持着分寸感和准确度。恰是这种分寸感，令空间的艺术性得以最大程度地呈现。

Lofter's features of relatively small size and flexibility fit into the current modern lifestyle and market structure. Height differences are cleverly used to integrate rather fragmented spaces so as to lay out the space. For spaces with compound functions, simple techniques should be adopted to ensure comfort home atmosphere and maintaining personality of the owner, instead of over labeling. These are the ways that help to get the big picture from small details through interior space design.

Simplicity does not mean no extra decoration. It is the essence of interior design to add proper decorations and furnishings which can express the owner's temperament. The apartment of Chongqing Poly Real Estate maintains a sense of propriety and accuracy both in space and decoration. It is this sense of propriety that presents the art of the space to the fullest.

利用建筑条件，巧妙布局不同的区间。动线自厨房－餐厅－会客厅－书房畅通无碍，开放的视线令空间充满活力和舒适。书房取代传统的电视墙，打破固有的空间功能分区，为客厅增添文化气息。额外增加的隔断解决了玄关处无遮挡的问题，也适当分隔开了餐厅与卫生间，保证舒适度；包裹其上的黑色烤漆面十分硬朗，为首层空间唯一的冷色调，因而更显时尚格调。

Building conditions are considered so as to cleverly lay out different sections. The unobstructed flow built an open view space, which makes it vibrant and comfortable. The study replaces the traditional TV wall, breaking the conventional spatial function division and adding a cultural atmosphere to the cube hall. And a partition is added to solve the problem of open hallway and to create spatial division. The black lacquer becomes the only cool color on the first floor, making it more fashionable.

❶❷ 客厅实景及效果图。
Rendering and picture of the living room.
❸ 从玄关眺望可看见首层的空间结构。
The hall way.
❹❺❼❽ 楼梯间一角。
The stairs.
❻ 首层平面图。
Plan of the first floor.

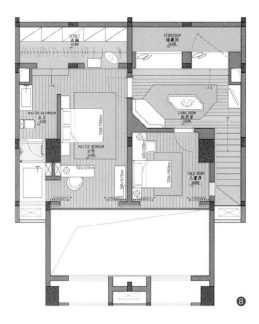

利用楼高以及对空间布局的娴熟掌握能力，二层划分了两个卧室和起居室。主卧及儿童房同享朝南的充沛光线，也可实现与楼下会客厅的交互。起居室是父母与子女共聚天伦的好地方。值得一提的是，主卧的衣帽间被设计为步入式开放结构。主人对潮流的把握和衣品的审美在橱窗静静展现，这也不失为一种自娱自乐的好方式。

Leveraging the height of the building and familiarity of space layout, the second floor is divided into two south-facing bedrooms and one living room. The sunlight interacts with the interior space and cause interacts with living room downstairs. The cloakroom in the main bedroom is designed with a walk-in open structure. The owner's fashion sense and dressing taste come into sight from the showcase, which is also a good self-entertaining way.

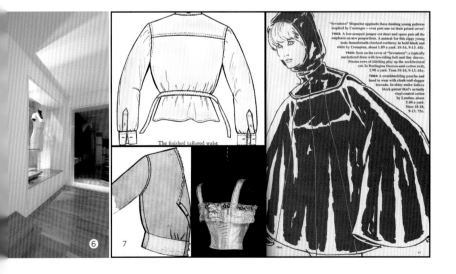

❶❷ 主卧套房。
The master room.
❸ 主卧洗手间。
The master bathroom.
❹❺ 二楼家庭厅。
The family room.
❻ 步入式衣帽间。
A walk-in cloakroom.
❼ 别具一格的陈设。
The details of furnishings.
❽ 二楼平面图。
Plan of the second floor.

合肥保利海上明悦销售中心
SALES CENTER OF POLY SEA BRIGHT, HEFEI

关键词：曲线、水袖枪花、水木相生
Keywords: arc-shaped lines, dancing sleeves and spears, balance between waters and trees

建筑面积 / 207,000 平方米
委托范围 / 硬装及软装设计
委托面积 / 689 平方米

曲线是建筑内外最直观的语言。舒展无拘束的线条赋予建筑格外灵动的基调。穿越千年的江月意象诉诸笔端，以现代简约的设计手法令建筑轻松将一池活水揽入怀中。在地的传统元素的糅合提炼，使设计的在地化变得更具识别度和亲切感。水木琳琅，相依相生，现代建筑中的徽派景致依旧有迹可循。

The stretched and free arc-shaped lines make the building livelier. The designer applies a modern and simple technique, introducing a pool of water into the building. Besides, the addition and refinement of local traditional elements promote the localization of designs, making them more recognizable and comfortable. With a balance between waters and trees, Hui-style can still be found in modern architecture.

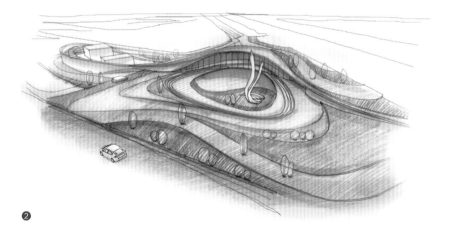

蜿蜒的整面落地玻璃如同取景器,让户外的水色如波浪般层层跌宕,天水一色的通透感令空间瞬间熠熠生辉。棕色木饰面所修饰的墙体既提供视觉上的依靠,又营造出傍岸的感觉。

The winding floor-to-ceiling glass is like a viewfinder, making the outdoor water rise like waves while the conjoining of water and sky makes the space shine instantly. The walls decorated with brown wooden veneers create both visual effects and a sense of being ashore.

❶❸❹ 弧形建筑包容在一片绿化广场之中。通高落地玻璃幕墙能为设计争取最大化的景观优势。
The arc-shaped building is included in a green square. The landscape advantage for the design can be maximized with the double-height floor-to-ceiling glass curtain wall.

❷ 建筑外观手绘图。
Sketch of the building appearance.

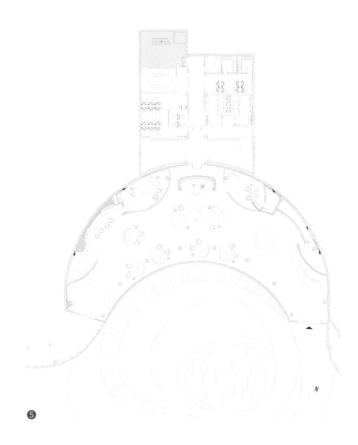

以暖灰色为主色调的现代家具配饰气质沉静,星罗棋布地点缀在空间中,如同水岸边被碧波打磨的鹅卵石。中岛式设计的水吧区增添了互动交流的乐趣,大片的LED光源令其成为整体空间最具时尚感的部分。

Modern furnishings in warm grey tones are embellished like stars in the space, resembling pebbles polished by green waves on the shore. The island-shaped bar area increases the fun of interaction as well as communication and becomes the most fashionable part with lots of LED light source installed.

❶❷ 柔软的内饰烘托氛围。
　　 Soft furnishings for foiling the atmosphere.
❸❹ 波浪形的立面。
　　 Wavy facade.
❺ 建筑平面图。
　　 Plan of the building.

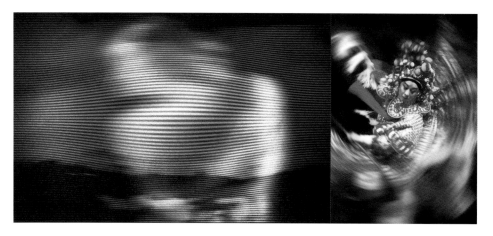
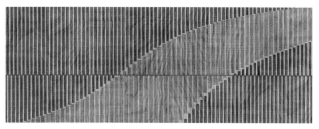
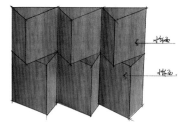

❶

庐剧曲调柔美，唱腔圆滑，本出自田间地头，是皖地重要的非物质文化遗产。适度提炼当地传统的文化元素，不但能帮助设计迅速切入正题，也为到访的客人营造了心理上亲切感。我们将庐剧中的水袖和枪花等元素化成繁密且律动成韵的木格栅，沿着建筑走势一路延展开去，营造出流动的美态。

Lu Opera, with soft melodies and smooth vocals, is an important intangible cultural heritage in Anhui. Moderate refinement of local traditional cultural elements not only ensures the design to the point, but also creates a psychological intimacy for guests. We apply the elements of dancing sleeves and spears in the opera into the dense and rhythmic wooden grilles, which stretch along the architecture and create a flowing beauty.

❶ 立面的推敲过程。
The evolution process of facades.
❷ 空间概念图。
Concept plan of the building.
❸❹ 静谧的氛围让人放松。
The quiet environment creates a relaxing atmosphere.

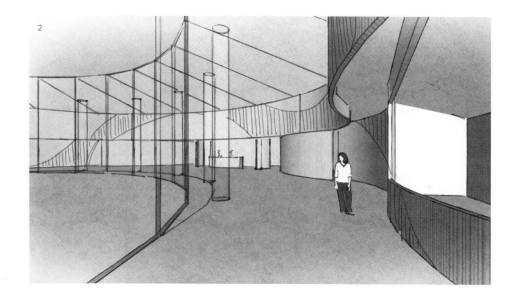
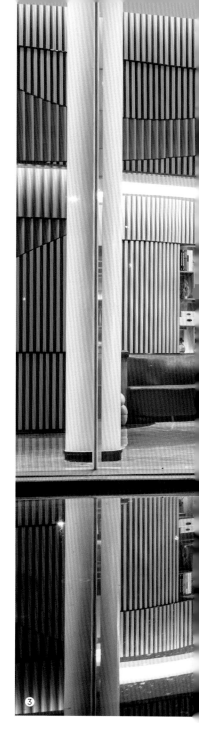

莆田保利拉菲公馆销售中心
SALES CENTER OF POLY LAFITE MANSION, PUTIAN

关键词：蓝色、简单、线条美
Keywords: blue, simple, beautiful lines

建筑设计 / 中国建筑集团公司
建筑面积 / 57,229 平方米
委托范围 / 硬装及软装设计
委托面积 / 320 平方米

海洋所带来的广袤想象投射到了空间中。喧闹以外的静谧之所，以简约的设计引海水之蓝加以修饰，重新赋予空间一个呼吸的节奏。保利莆田拉菲公馆的整体设计化繁为简，尽显精致与细腻。我们运用时尚的线条，以清透、顺畅、简洁为原则，排除视觉的干扰，令空间布局与使用功能完美结合。在黑白相间的大背景下，时尚明快的蓝色调和质感巧妙地平衡了空间色彩的黑白中性基调，在同一空间中完成了温暖与冰冷的对话。

Poly Lafite adopts a simple design with sea blue decorations, presenting an exquisite and delicate style. Followed by the principles of clearness, smoothness and neatness, we use fashionable lines to remove visual interference and realize a perfect combination between space layouts and functions. In the black and white background, the stylish blue tone and texture subtly balance the black and white neutral tone, an encounter of warm and cold.

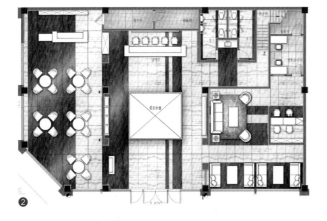

空间中的黑白灰三色过渡自然，模糊了色彩的边界，跳跃而出的宝蓝色十分亮眼。接待前台及沙盘区依进深排列，灯饰造型中的多处线条的延展性彰显空间的深远，点晕开般的大理石纹路源于自然的造化。在挂画及绿植点缀下，浓郁的自然艺术气息令人浮想联翩。

With black, white and gray as the main tone, the prominent royal blue stands out in the space. The reception desk and the sand table are arranged successively. The designer uses stretched lines in lighting to reflect the depth and wideness of the space and marbles with ripple patterns to deliver a natural air. In addition, paintings and green plants add a rich and natural artistic atmosphere.

❶ 宽敞的沙盘区。
Spacious sand table area.
❷ 销售中心平面图。
Plan of the sale center.
❸ 造型独特的吊灯。
Unique Chandelier.
❹❺ 宝蓝色点缀着空间。
Royal blue decorated around the space.

❶

❷

VIP洽谈区中缎面的宝石蓝和苍翠的绿意延续其中,夺人眼球。宝石蓝的沙发新颖特别,就像是深邃的海洋之眼。坐落于此,客人可自由对话,轻松自在。简单的线条及雅致的颜色组合,加上现代元素的灯饰家居,让空间既舒适又宽阔。空间本身充满想象,自然也就能让周遭的人获得更多灵感。

In the VIP negotiation area, the satin-faced sapphire blue and verdant green are eye-catching; the sapphire blue sofa is novel and special, like the eye of the sea. Sitting here, guests are free to talk and feel at ease. The combination of simple lines and elegant colors, coupled with modern lighting, makes the space comfortable and spacious. The space itself is full of imagination and accordingly give inspirations to people here.

❶❷ VIP洽谈区的实景图及效果图。
Rendering and picture of the VIP area.
❸❹ 绿色的植物与宝蓝色搭配和谐。
Green plants with royal blue.
❺ 水吧区。
The bar.

北京和锦薇棠销售中心
SALES CENTER OF POLY BELLE VILLE, BEIJING

关键词：去售楼部化、艺术、阶梯
Keywords: de-marketing, art, stairs

建筑设计 / 北京市住宅建筑设计研究院亚欧西安公司
建筑面积 / 117,000 平方米
委托范围 / 硬装及软装设计
委托面积 / 1,059.18 平方米

北京CBD东扩线上的热土备受瞩目。保利地产在此建造属于未来的现代艺术之所。秉承保利品牌基因中的"和"筑善，以"锦"寓意为居者构筑美好，取义"薇"寓意都市的美好年华，用现代艺术连通项目的景观、建筑与设计，让室内外的景观相通相融。对于现代消费者而言，一个符合他们生活品味追求的空间更能吸引他们的眼球。室内空间的设计除了有基本的舒适感之外，也要富有个性和张力以满足他们的品味的追求。

The land on the east expansion of Beijing CBD attracts tremendous attention, where Poly Real Estate builds promising residences of modern art. Its Chinese name "Hejinweitang" carries significant meanings: "He" (harmony) is a brand legacy; "Jin" (gorgeous) means to deliver happiness to residents; "Wei" (common vetch) means good times of the city. The application of interconnection among landscapes, buildings and designs integrates the indoor and outdoor landscapes. For modern consumers, a space in line with their pursuits for quality life is more eye-catching. To this end, besides being cozy, interior designs should also be characteristic and attractive.

为了凸显整体空间的质感，我们采用相对细腻的马来漆作为装饰涂料。首层以阶梯式活动区为中心。宽面的阶梯可作为阅读区或艺术品陈列区，多功能的设计呈现出前卫与实用共生的先锋理念。

To highlight the texture of the whole space, we apply the relatively smooth Malay paint. The first floor is centered by a multistep functional area, where the wide stairs can turn into reading area or exhibition area. The multifunctional design displays a pioneering concept—the coexistence of fashion and practicality.

❶ 建筑平面图。
Plan of the building.

❷❸❹❺ 以空间作为展示物的销售中心,将解构主义手法延伸室内,以简洁、重构作为设计的核心注入空间,打造艺术与当代,前卫与实用共生的先锋设计。
With the space as the sales center of the display, the deconstruction approach is extended to the interior with simplicity and reconstruction as the design core for injecting into the space, so as to create a pioneered space where art and the contemporary era as well as the avant-garde and practicability coexist.

二层空间中,与阶梯相连的水吧选用锦玉的大理石材搭配现代吊灯,彰显现代艺术装饰的同时,亦寓意着来访者更上一层楼的美好生活。休闲区与VIP接待室,以冷色系中的灰色调作为软装配饰的主色调,用些许绿植衬托空间,简约的装饰点缀以突出空间设计,是设计师对艺术生活的理解。

The bar connected with the stairs on the second floor applies jade marble coupled with modern chandelier, presenting modern art and meanwhile wishing visitors a happy life. The leisure area and VIP reception room use grey as the main tone of soft-packed decorations, accompanied by some green plants. Such simple decorations highlight the space design and demonstrate the designer's understanding of artistic life.

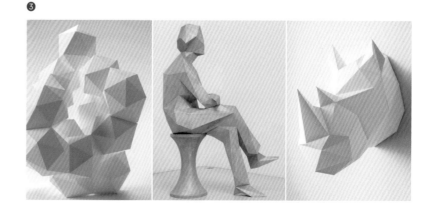

❶❹ 二层VIP洽谈室。
The VIP area on the second floor.
❷ 二层平面图。
Second floor plan.
❸❺ 清新可爱的陈设。
Fresh and lovely furnishings.
❻❼❽ 空间内部的光影之美。
The art of light and shadow of spaces.

现代主义的实用与大气,来自于去芜存菁的胆识,将美的那部分表达得极致又恰如其分。售楼中心作为住宅项目的门户,一向承载着捕捉注意力、增强楼盘辨识度的期望,因此我们尝试以现代主义的艺术性为突破口,统一空间内外的设计手法,令一处建筑化成一个"纯粹"的符号,一座柏拉图式的极简主义空间由此而生。外建筑与内建筑空间矗立在葱茏的树影中,如同绿荫草坪上的一处耀眼的白色山丘。

Sales center, as the portal to the housing projects, has always been an eye-catcher, helping to make the projects more recognizable. Therefore, we attempt to apply modern art and unify the exterior and interior designs, turning the whole estate into a "pure" symbol—a platonic minimalist space. The whole estate stands in the shadows of luxuriant trees, just like a bright white hill on a verdant lawn.

高达4米的高通透性隔热玻璃幕墙取代了以往阻隔视线的墙面，环绕整个建筑，室内室外俨然成为一个无界状态。地面的的原石材料从前庭延伸到空间内部，寓意着自然无界。入口处一块原始、纯粹的石头经过设计的巧工雕凿后，其一化为接待处，其二化为室外的休息区，既保留了自身的形态，又在这个纯粹的空间中，呈现出不同的功能形态和艺术感。

A 4-meter-high glass curtain wall with high transparency and thermal insulation surrounds the whole building replacing the conventional solid wall, making the indoor and outdoor seemingly boundless. The raw stone material extends from the courtyard to the interior, implying a boundless and consistent natural style. A raw stone at the entrance is carefully carved and turned into a reception and a leisure area at the outside, retaining its own form while with different functions and aesthetic feelings.

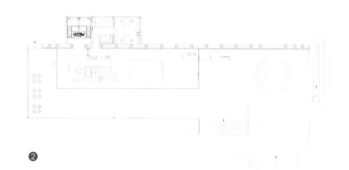

❶ 近四十米建筑面宽的纯粹"盒"式建筑体中没有一梁一柱，这对于建筑设计是难题，对内建筑设计同样是一个挑战。
There is neither a beam nor a column in the "box" building with a width of nearly 40 meters, which is not only a difficulty for building design, but also a challenge for interior building design.
❷ 建筑平面图。
Plan of the building.
❸❹ 石材寓意着自然纯粹。
Stone means purity and nature.

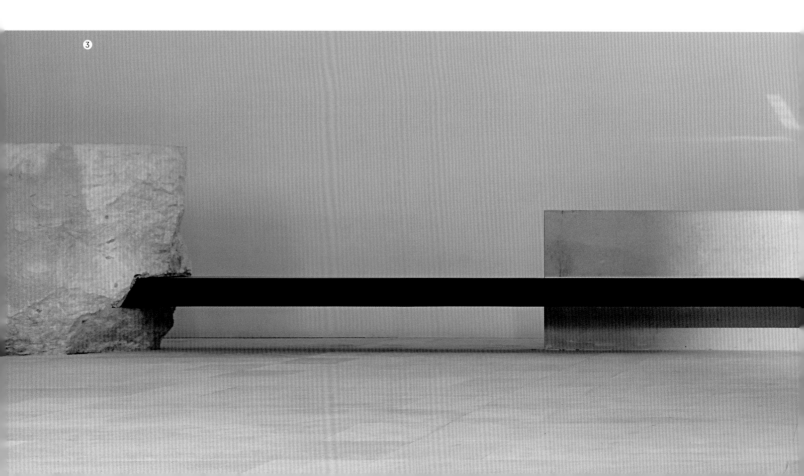

❶❸❹ 建筑没有墙、柱、转角的掩蔽，搭配近四米的通高玻璃幕墙，整体空间轻盈平衡。因此设计师选用简约的家具以呼应这种特别的纯粹感：横平竖直的家具体态，搭配柔软舒适的材质，令人放松平静。
Without any walls, columns, and corners, the building is equipped with a full-height glass curtain wall in nearly four meters, contributing to a light and balanced space as a whole. Therefore, a simple set of furniture is selected by the designer to echo to the special sense of purity: the horizontal and vertical furniture details enables peoples to feel relaxed and peaceful with soft and comfortable materials.

❷ 建筑外观手绘图。
Sketch of the building appearance.

在这里，没有任何一件附加于建筑之上的多余之物，没有杂乱的装饰，有的只是轻盈通透的空间感受和内外流动的阳光空气。一个适合会谈交流的"城市客厅"，正将生活化的场景延续到公共空间。

Here, there is no redundant elements or messy decorations, but a light and bright space with full convection of air and light. An "urban living room" for talks and exchanges extends the life-based setting to public space.

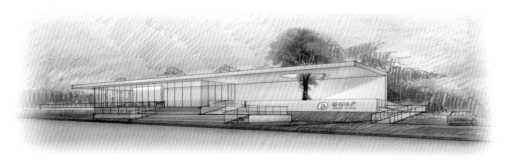

❶ 户内户外，若即若离的乌托邦。
A cozy and relaxing Utopia.
❹ 优雅的陈设。
It's full of elegant furnishings.

现代主义的艺术表达中，建筑、景观、室内装饰，无不在低调地点题。设计的统一性和完整性，使空间形态、视觉元素、应用材料乃至功能分区，都充分呼应了周边环境，让室内室外、动与静互动达到一个巧妙的平衡——以极简映衬繁华，构筑一个舒适放松，与现实若即若离的乌托邦。

In the artistic expressions of modernism, buildings, landscapes, interior decorations all highlight the theme in their own way. The uniformity and integrity of the design allow the space structure, visual effects, material and functional division to echo their surroundings, creating a subtle equilibrium between the indoor and outdoor as well as the motion and stillness. The luxurious design in a minimalist style builds a cozy and relaxing Utopia at an arm's length from the reality.

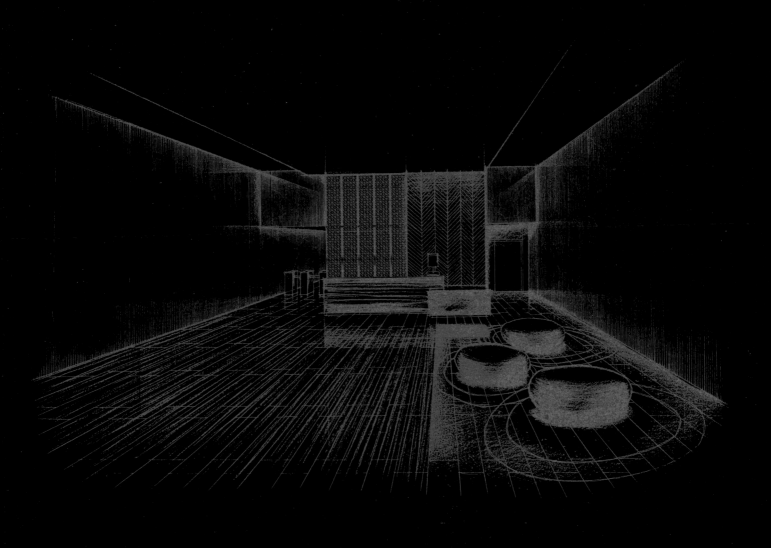

源自生活，追求卓越

从2008年到2018年，我们终于迎来了创业的第一个十年。这是我们从零到壹的十年，是我们人生事业第一阶段的逗号，也是我们总结自我、整装待发，再次起航迈向未来十年的元年。

过去十年，我们时刻地审视自己：有没有把当年在广州美术学院念书时，众多的学界泰斗，诸如尹定邦老师、赵健老师、吴卫光老师、沈康老师、杨一丁老师、林红老师、杨岩老师、金涛老师，还有校办企业集美组设计公司的林学明老师、陈向京老师、梁建国老师、张宁老师、齐胜利老师、蔡文齐老师等前辈的谆谆教导和教诲寄以言行？同时我们学着屏蔽世间浮华，潜心治学和求证，在探索设计的路上执着地自我修行和反省：怎样才能实现设计师的自我价值，我们的人生追求目标是什么，我们应该保持什么样的职业态度，传承前人的智慧和经验；我们应该为社会做出怎样的贡献，向年轻的下一代的设计师树立榜样，令他们有坚定的信念，让设计的理念得以薪火相传。

我和王小锋毕业于广州美术学院环境艺术设计专业，在校的时候接受的设计训练并非只有绘画，还有建筑初步、现代建筑设计史论等系统化、标准化的建筑学训练。为此我要感谢我们的班主任沈康教授和杨一丁教授，他们都是非常严谨治学的建筑师。学习建筑设计的史论部分让我们明白了传承前人的智慧和寻找不同学派脉络基因的重要性；学习建筑史、设计方法论容不得半点形式上的虚假，必须以严谨的辩证思维求证。以上的一切，决定了由我们所在的广州尚诺柏纳空间策划联合事务所（以下简称SNP）是治学严谨、客观踏实、有设计脉络并且有责任感的专业团体。

第二部合辑是SNP近5年以来思想和行动统一的结晶。这次我们一改以时间轴为主线的编撰思路，把我们的设计业务进行分门别类，并挑选代表性的项目与之对应。此外还有很多非常精彩的设计构思，由于编撰精力有限而尚未能汇编入册，颇有些可惜。但正如我们曾经强调的，如何分析和看待每一个项目的特征并加以解决问题，才是设计的立命之本。至于风格问题，我们认为更多是视觉呈现和感官触觉方面的处理。

"探索设计中的生活，实践生活中的设计。"一切社会生产活动的本源，正是为了满足人类生活的孜孜追求。"源自生活，追求卓越"这八字座右铭，是我们经营哲学的由来和精髓概括。我们仍在校读书时，曾阅读刊登在《世界建筑导报》里关于香港大学建筑系教育的专题，文章详细地阐述了其对香港人100多年来的生活研究。香港的设计教育始终关注怎么解决本土的人居生活矛盾，这给了我们国内的同行们最直接的启示和借鉴作用。香港巴马丹拿建筑事务所跨越150年历史，培养了超过6代建筑师，为香港设计薪火相传作出了卓越的贡献。其务实、经典的作品，包括上世纪30年代的上海和平饭店及今日的香港渣打银行，仍然是历久弥新的建筑典范，在在城市的记忆里、人们的生活里留下了不可磨灭的印记。"关注民生，解决社会问题，推动社会生活发展"，这无疑是值得我们所有设计人传承的态度。

包豪斯设计学院（BAUHAUS）作为现代设计教育的典范，它所提倡的艺术与技术统一的思想，以及为设计教学与实践结合的模式推动了现代设计的发展。我们秉承其的治学理念，用设计专业回馈社会，将SNP塑造成一所朝气蓬勃，坚持平等、民主、互惠的社会实践基地。我们始终为能与华南地区学府派出身的一群同窗好友共同奋斗而感到无比的幸福和自豪。三人行必有我师，衷心感谢为此书作序的两位良师益友。他们分别是：文化学者关鸣先生；1998年的《巴马丹拿130年》作品集编撰者、推行者；以及羊城设计联盟会长、SPDG汉森伯盛国际设计集团的创始人盛宇宏先生，他是中国早期建筑设计行业中注册民营设计院的先行者之一，可谓领先一时风气。

最后，谨以此记，纪念曾经在华南大地上勇于创新和实践的人们，也提醒自己把前人递过来的薪火传承下去，更鼓励后人发愤图强、不忘初心。与诸君共勉。

FROM LIFE TO LIFE

From 2008 to 2018, this year of 2018 marks the first decade of our entrepreneurship, from zero to one. It is the comma at the first stage of our career as well as the beginning for us to review and get ready to set sail again.

For the last decade, we have constantly questioned ourselves: Have we put those earnest teachings into practice? Teachings were from the famous scholars at Guangzhou Academy of Fine Arts like Mr. Yin Dingbang, Mr. Zhao Jian, Mr. Wu Weiguang, Mr. Shen Kang, Mr. Yang Yiding, Ms. Yang Hong, Mr. Yang Yan, and Mr. Jin Tao as well as the teachers from the school-run firm Newsdays like Mr. Lin Xueming, Mr. Chen Xiangjing, Mr. Liang Jianguo, Ms. Zhang Ning, Mr. Qi Shengli, and Mr. Cai Wenqi. Meanwhile, we have focused on the design study and repeatedly reflected on ourselves: How to realize a designer's self-value? What do we pursue for? What professional attitudes should we keep? How do we carry on the predecessors' wisdom and experience? What contributions should we make to the society? How do we set an example for younger designers to empower them and pass down the design philosophy from generation to generation?

Wang Xiaofeng and I, majored in Environmental Art Design and graduated from Guangzhou Academy of Fine Arts. We received systematic and standard architecture training like Elementary Architecture as well as History and Theory of Modern Architecture, not just painting. Here I want to thank Professor Shen Kang and Professor Yang Yiding, both with conscientious scholarship. The study of History and Theory of Modern Architecture has helped me to understand it's important to inherit the predecessors' wisdom and figure out the root of every different schools of design. And the study of its history and methodology asks for 100% scrupulous dialectical thinking. And all of these have contributued Sunny Nuehaus Partnership (SNP) to be a responsible and professional organization with conscientious attitude and notable design achievements.

The second collection gathers works of SNP for the past 5 years. This time we categorize designs into different types and select according representatives, instead of using timelines. And it is a pity that there are many unlisted fantastic design cases due to the lack of time and energy. But as what we have emphasized, it is the root of design to carefully analyze every individual project and solve the problem accordingly. As for style, it is more about visual effect and sensations.

Derive from life and strive for excellence" explains our business philosophy. When we were at school, I read a feature about the education of the Architecture Department of Hong Kong University on Architectural Worlds, which conducted a detailed study of Hong Kong (HK) people's life for the past century. HK's design education always focuses on how to solve the problem between residence and humanity, which enlightens us most as mainland counterparts. HK P&T Group, with a history of 150 years, have trained 6 generations of designers and made great contributions to HK designs. Its pragmatic and classic works including Shanghai Fairmont Peace Hotel in the 30s and today's Standard Chartered in HK are everlasting architectural paradigms, imprinted in the memories of the people and the city. "Focus on people's livelihood, solve social problems and promote social development"—such attitudes are what we designers should inherit.

BAUHAUS, as a vanguard of modern design education, advocates the unity of art and technology as well as the mode of combining teaching and practicing, which promotes the development of modern design. We carry on its ideas—to give back to the society with our professional designs, and turn SNP into a promising social practice base, upholding equality, democracy and mutual benefits. It has always been a pleasure and privilege to work with schoolmates and friends from the same academic circle in southern China. "Two heads are always better than one". My heartfelt thanks go to my two dear friends who write prefaces for this collection: famous scholar Mr. Tom Kwan, compiler and promoter of *P&T Group: a History of 130 Years* (1998); and Mr. Sheng Yuhong, President of Guangzhou Design Alliance, founder of SPDG, and one of the pioneers of registering private designing institute in the early Chinese architecture industry.

Last but not least, something I would like to share with you all: commemorate those creative and pragmatic vanguards in southern China, remind myself of passing on the torch, and encourage the younger generation to strive forward.

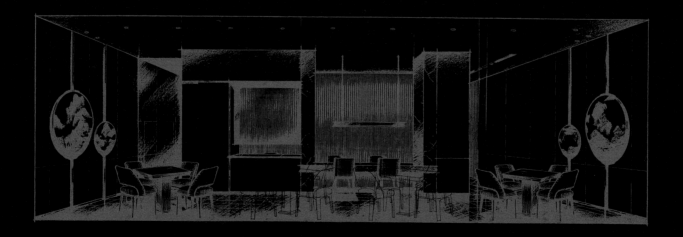

从零到壹的视界

2018年对我和王小峰来说,是五味杂陈又意义非凡的一年。这是一段历史的结束,也是另一段未知旅途的开端,箇中滋味难以言表。十年后的深夜里,我们提笔写下这些文字,过去的林林总总,历历在目。

2008年8月8号,SNP搬进了广州信和商务中心,130㎡的单元,创始人4名,学徒3名,年底一桌饭,和睦一家人。然而两个月后,金融风暴席卷全球,我们只能一边学习一边思考,内心难免焦灼。

2009年的经济依然不景气,但生活仍要继续。王赟往返北京和广州,王小锋新婚不久,许涛开拓业务,于玲也重新投入职场。2010年春天,赖艺超和王赟的爱徒周嘉妍先后加入公司。这一年的尾牙宴上,于玲高兴地宣布,本寺18铜人,化缘2400万。

事情终于慢慢步入了正轨。2011年初秋,SNP搬入广州白云路白云大厦。比邻原广九铁路广州火车站。中国大陆开往香港的第一列火车,就是从这儿开出。此时,记42人,规模初显。

2011年底,广州城改史上的最重要一笔旧城改造项目——琶洲新城改造拉开序幕,我们有幸参与其中,项目开发总计185万平方米,用时超过6年。由浅入深,投入重兵,培养起了众多新一代设计力量。

2013年6月,SNP承接三亚国家财经论坛中心主体超高层建筑——海悦公馆的室内设计。2017年,该建筑交付使用,发挥其财经政策交流作用。与此同时,SNP启动了企业内部管理改革。广州美术学院60周年大典之际,经美术学院沈康教授推荐,王赟进入优秀校友名录,系环境艺术设计专业自1990年正式设立以来获得此殊荣的8位校友之一。同年,广东省陈设艺术协会成立,王赟担任协会第一届副会长和常务理事。在此,也要感谢执行会长胡小梅女士、秘书长梁欢先生的建议和指点。

2014年,SNP已经拥有超过120名设计师,王小锋为公司争取到参与厦门保利叁仟栋大型海滨项目的室内设计的机会。同年,由知名室内设计师谢英凯先生及好友韦杰推荐正式加入广州民营设计企业的联盟——羊城设计联盟。

2015年,SNP向逾十家亚欧顶尖的设计企业学习设计、管理经验,加快改革步伐,往中心制及事业部制的企业架构靠拢。截止2015年年底,三大管理中心已具雏形。

2016年春季,SNP再次迁至天河区天伦控股大厦,临近广州东站。这一年,承接逾10栋超高层建筑的室内设计,包括著名的琶洲天幕广场,业绩突破亿元大关。同时我们重新思量企业在社会中的角色,积极开展校企合作以回馈社会,培养设计新秀。

2017年,SNP完成股东会和董事会的调整及分配。王赟担任董事长和总裁,王小锋担任总设计师,许涛担任监事和经营管理中心总经理,于玲担任运营管理中心总经理。公司聚集超过200名设计师,开始产业链跨界与整合,把"格物致知,知行合一,修心明志,传承创新"定为企业格言。同年,顺利完成国内某银行接待中心、知名企业位于沪、渝、鹏的企业会馆或大型科技园展厅,某私人博物馆等特殊的设计任务。

迈进第10个年头,SNP确立了"传承文化和认真思考"乃是设计咨询行业第一价值的发展方向,正式把"产学研"的经营模式提上未来的发展议程,以"专题研究所"和"学术顾问"双轨并行的发展计划,积极探索下一阶段的方向。就在落笔撰文时,王小锋开始尝试珠海保利大剧院的文创空间策划案;天津南开大学地块的旧改项目也正由王赟主导推进。这两个大型公共空间的室内设计项目,相信将会成为两位步入不惑之年的两位设计师在职业生涯上的精彩一笔。

至此,我们真诚地感谢为这本书能够顺利出版面世而提供了宝贵时间、精力的好友同事们。他们是好友康建国先生,廖荣辉先生,同事杨棉武先生,张文伟先生,陈嘉琳女士,罗晓明女士,麦港飞女士,黄早慧先生,罗彩甜女士等等。参与的人员众多,在此不一一赘述,谢谢大家。

再次感谢关鸣先生和盛宇宏先生在百忙之中为此书立序。籍本书发行之际,十二万分感谢栽培我们设计慧根的沈康老师和杨一丁老师。最后,我们一并衷心感谢我们的家人,是他们默默地在背后支持我们的事业,照料我们的生活,才让我们能在人生的道路上执着自信地前进,开始我们从1到2的旅程。

后记 / POSTSCRIPT

FROM ZERO TO ONE

The year 2018 is of extraordinary significance for me and Wang Xiaofeng. It is the end of a journey, yet the beginning of another unknown one. It's hard to express our feelings. At the night reflecting on the past decade, we wrote down these words as if the past was in sight.

On August 8, 2008, SNP moved into a 130m2 office at Guangzhou Xinhe Business Center. 4 founders, 3 apprentices, a dinner at the end of the year, we were like a family at that time. However, two months later, the financial storm swept the world. Be it anxious, we could only keep studying and thinking.

The economy was still sluggish in 2009, but we still had to go on with our life. Wang Yun traveled to and fro between Beijing and Guangzhou; Wang Xiaofeng was newly married; Xu Tao opened up new businesses; and Yu Ling rejoined the workplace. In the spring of 2010, Lai Yichao and Zhou Jiayan (the apprentice of Wang Yun) joined the firm. At the end of the year, Yu Ling was pleased to announce that the firm, with 18 staff, got an investment of 24 million yuan.

Things were gradually on the right track. In the early autumn of 2011, SNP moved into the Baiyun Building in Baiyun Road, Guangzhou, adjacent to Guangzhou Railway Station located at the original Guangzhou and Kowloon Railway. The first train from mainland China to Hong Kong was from here. At that time, SNP had 42 staff, and the scale was first revealed.

At the end of 2011, the most important old city reconstruction project in the history of Guangzhou -- the transformation of Pazhou New City, kicked off. We were fortunate to be part of it.

The project totaled 1.85 million square meters and took more than 6 years. As time went on, the project accumulatively cultivated quite a new batch of design talents.

In June 2013, SNP undertook the interior design of Haiyue Mansion, the main super high-rise building of the Sanya National Financial Forum Center. In 2017, the building was put into service and played its role of financial policy exchange. At the same time, SNP initiated an internal management reform. On the occasion of the 60th anniversary of Guangzhou Academy of Fine Arts, recommended by Professor Shen Kang of the Academy, Wang Yun joined the list of outstanding alumni, one of the eight alumni who received this honor since the establishment of the Environmental Art Design in 1990. In the same year, Guangdong Association of Art Crafts and Decoration Industry was established, and Wang Yun served as the first Vice President and Executive Director of the Association. Here, I would also like to thank the Executive President Ms. Hu Xiaomei and the Secretary-General Mr. Liang Huan for their opinions and suggestions.

In 2014, SNP had more than 120 designers, and Wang Xiaofeng won the opportunity for the firm to participate in the interior design of the large-scale waterfront project of Poly Coastal Mansion in Xiamen. In the same year, recommended by the well-known interior designer Mr. Xie Yingkai and his friend Wei Jie, SNP formally joined the alliance of Guangzhou private design firms - Guangzhou Design Alliance.

In 2015, SNP learned design and management experience from more than ten top design firms in Asia and Europe, accelerated the pace of reform, and taken the enterprise structure of the central system and divisional system. By the end of 2015, the three major management divisions were formed.

In the spring of 2016, SNP moved to Tianlun Holdings Building in Tianhe District, adjacent to Guangzhou East Railway Station. This year, SNP undertook the interior designs of more than 10 super high-rise buildings, including the famous Poly Skyline Plaza in Pazhou, and its business volume exceeded 100 million yuan. At the same time, we reconsidered the role of enterprises in society, and actively carried out university-enterprise cooperation to give back to the society and developed up-and-coming design talents.

In 2017, SNP completed the adjustment and distribution of the board of shareholders and the board of directors. Wang Yun served as Chairman and President, Wang Xiaofeng Chief Designer, Xu Tao Supervisor and General Manager of the Business Management Center and Yu Ling General Manager of the Operation Management Center. The firm gathered more than 200 designers and started the cross-border integration of the industrial chain. It has set the corporate motto as "the pursuit of knowledge, the unity of knowledge and action, the cultivation of mind, the inheritance and innovation". In the same year, it successfully completed a series of challenging design projects, including a domestic bank reception center, office buildings of a well-known enterprise in Shanghai, Chongqing and Shenzhen, the exhibition hall of a large science and technology park, and a private museum.

In the 10th year, SNP established "culture inheritance and serious thinking" as the development direction of the design consultation industry and officially put the business model of "industry-university-research cooperation" on the future development agenda. Following the dual-track development direction of "special research institute" and the "academic advisor", SNP actively started to explore the direction of next stage. As of writing the article, Wang Xiaofeng was actively planning the creative space scheme of the Poly Theater in Zhuhai while Wang Yun was in charge of the old land reform project of Nankai University in Tianjin. The design projects of the two large public spaces are believed to be a milestone for the two designers at the age of forties.

Hereby, we sincerely appreciate our friends and colleagues who have spent valuable time and energy for the publication of this collection, including our friends Mr. Kang Jianguo and Mr. Liao Ronghui, colleagues Mr. Yang Mianwu, Mr. Zhang Wenwei, Ms. Chen Jialin, Ms. Luo Xiaoming, Ms. Mai Gangfei, Mr. Huang Zaohui, Ms. Luo Caitian and so on. The list can go on and on... In a word, thank you all.

Thanks again to Mr. Tom Kwan and Mr. Sheng Yuhong for taking time to write prefaces for this collection. On the publication of this collection, my heartfelt thanks go to Professor Shen Kang and Professor Yang Yiding for their enlightenment and cultivation. Finally, we sincerely thank our family members for their full support on our career and life, so that we can march on confidently and begin our journey from One to Two.

图书在版编目（CIP）数据

研行十载：SNP实践与回顾.第二辑 / 广州尚逸装饰设计有限公司编著.—北京：中国林业出版社,2018.11

ISBN 978-7-5038-9847-1

Ⅰ.①研… Ⅱ.①广… Ⅲ.①室内装饰设计-作品集-中国-现代 Ⅳ.①TU238.2

中国版本图书馆CIP数据核字(2018)第261721号

研行十载——SNP 实践与回顾 第二辑

策划	广州市唐艺网络科技有限公司
编著	广州尚逸装饰设计有限公司

策划编辑	高雪梅
策划指导	王　赟　杨棉武
文字整理	麦港飞　罗晓明
英文翻译	罗彩甜
手稿绘制	刘卓海
装帧设计	廖荣辉
图片摄影	黄早慧

中国林业出版社·建筑分社

责任编辑	纪　亮　王思源

出版发行	中国林业出版社
出版社地址	北京西城区德内大街刘海胡同7号，邮编：100009
出版社网址	http://lycb.forestry.gov.cn/

印刷	恒美印务（广州）有限公司
开本	1016mm×1320mm 1/16
印张	24
版次	2018年12月第1版
印次	2018年12月第1次
定价	338.00元

图书如有印装质量问题，可随时向印刷厂调换（电话:020-84981812)